UNDERSTAND, MANAGE, AND PREVENT ALGORITHMIC BIAS

A GUIDE FOR BUSINESS USERS AND DATA SCIENTISTS

Tobias Baer

Apress®

Understand, Manage, and Prevent Algorithmic Bias: A Guide for Business Users and Data Scientists

Tobias Baer
Kaufbeuren, Germany

ISBN-13 (pbk): 978-1-4842-4884-3   ISBN-13 (electronic): 978-1-4842-4885-0
https://doi.org/10.1007/978-1-4842-4885-0

Managing Director, Apress Media LLC: Welmoed Spahr
Acquisitions Editor: Shiva Ramachandran
Development Editor: Laura Berendson
Coordinating Editor: Rita Fernando

Cover designed by eStudioCalamar

Distributed to the book trade worldwide by Springer Science+Business Media New York, 233 Spring Street, 6th Floor, New York, NY 10013. Phone 1-800-SPRINGER, fax (201) 348-4505, e-mail orders-ny@springer-sbm.com, or visit www.springeronline.com. Apress Media, LLC is a California LLC and the sole member (owner) is Springer Science + Business Media Finance Inc (SSBM Finance Inc). SSBM Finance Inc is a **Delaware** corporation.

For information on translations, please e-mail rights@apress.com, or visit www.apress. com/rights-permissions.

Apress titles may be purchased in bulk for academic, corporate, or promotional use. eBook versions and licenses are also available for most titles. For more information, reference our Print and eBook Bulk Sales web page at www.apress.com/bulk-sales.

Any source code or other supplementary material referenced by the author in this book is available to readers on GitHub via the book's product page, located at www.apress. com/9781484248843. For more detailed information, please visit www.apress.com/ source-code.

Printed on acid-free paper

*For the love algorithm in my partner's mind—
I still haven't figured out what bias caused him
to choose me but I think it's the best mistake
he has made in his life!*

Contents

About the Author

Tobias Baer is a data scientist, psychologist, and top management consultant with over 20 years of experience in risk analytics. Until June 2018, he was Master Expert and Partner at McKinsey & Co., Inc., where he built McKinsey's Risk Advanced Analytics Center of Competence in India in 2004, led the Credit Risk Advanced Analytics Service Line globally, and served clients in over 50 countries on topics such as the development of analytical decision models for credit underwriting, insurance pricing, and tax enforcement, as well as debiasing decisions. Tobias has been pursuing a research agenda around analytics and decision making both at McKinsey (e.g., on debiasing judgmental decisions and on leveraging machine learning to develop highly transparent predictive models) and at University of Cambridge, UK (e.g., the effect of mental fatigue on decision bias).

Tobias holds a PhD in Finance from University of Frankfurt, an MPhil in Psychology from University of Cambridge, an MA in Economics from UWM, and has done undergraduate studies in Business Administration and Law at University of Giessen. He started publishing as a teenager, writing about programming tricks for the Commodore C64 home computer in a German software magazine, and now blogs regularly on his LinkedIn page, www.linkedin.com/in/tobiasbaer/.

Acknowledgments

First of all, I want to thank my publisher, Shiva Ramachandran, who deserves sole credit for coming up with the brilliant idea of writing this book, and my editor, Rita Fernando, who not only unleashed a writing beast in myself through her relentless encouragement but also kept the red ink away from my quirky humor. She's to blame for everything—I had expected a lot more adult supervision!

I also want to thank Professor (now emeritus) Paul Shaman from the Statistics Department of The Wharton School of University of Pennsylvania with whom I had the privilege to spend two precious months in 1999 as a Visiting Scholar. He opened my eyes to the difference between running a script to estimate a model and understanding data—much of my critical attitude towards data originates in his teachings.

I finally would like to thank Clemens Baader, who graciously read the draft manuscript and was always a fabulous sounding board for my ideas.

Preface

Why did I write this book? Much has been written about algorithmic biases already; disturbing examples of algorithmic biases abound. Much less has been written about the actual causes of algorithmic biases, however, and very little seems to be known about how to solve the problem and either prevent algorithmic biases altogether or manage them in a way that prevents harm. This is what this book is about.

This book is practical. It suggests solutions that you can start implementing tomorrow. Some of the actions may take some time to reach completion or fruition—but this book is not about fancy theory. There are step-by-step guides and check-lists, as well as countless real-life examples to illustrate my points. Most importantly, though, this book encourages critical thinking by suggesting which specific questions to ask.

The more I discovered about algorithmic biases in my own modeling and consulting work, the more I realized that it is much more than a technical issue. Yes, statistics accounts for both some of the root causes of algorithmic biases and some of the solutions. However, the issue is deeply rooted in human psychology, and we cannot address algorithmic biases without understanding human biases and how the biases of users, data scientists, and society at large create and proliferate decision biases.

Therefore I do not jump right into technical solutions but take the time to explain where algorithmic biases come from—and what this means for fighting them.

And the (non-technical) users of algorithms—such as business managers and public servants—have a lot more power to fight and prevent algorithmic biases than they might believe. This book wants to empower everyone to better deal with algorithmic bias and join hands to prevent it.

Who This Book Is For

We live in a world where all of us are affected by algorithms and many of us use them, maybe even unaware that an algorithm is involved. Therefore I have written this book for all of us.

Data scientists are the scarce experts who develop algorithms and therefore have a big role to play in dealing with and preventing algorithmic bias, and I therefore hope that many, maybe all of them, will read this book—and the last, most technical part of this book is even dedicated to them.

However, most people are not data scientists, and many outright hate statistics. The book therefore is written with a lay(wo)man in mind. It uses nontechnical language, vivid analogies, and tries to keep the fun quotient up through excessive use of humor. Warning: You might even like statistics in the end, at least the biased image of it projected by this book...

As the issue of algorithmic bias has come to the fore, naturally also compliance officials and regulators have started to explore it and search for ways to prevent harm from it. Therefore, this book is not just for the actual developers and users of algorithms—and the business managers and public servants who decide where and how to use them—but also for compliance officials and regulators tasked with keeping decision processes in check.

And, as I will argue, many algorithmic biases are a mirror of deep-rooted societal biases. Therefore the issue of algorithmic biases is a much larger one, and I have written this book also for politicians, journalists, and philosophers who need to know that algorithms can be as much a solution for fighting societal biases as they can be a problem if they perpetuate and amplify such biases.

Last but not least, the book is for Martians and Zeta Reticulans. You'll figure out why soon!

What This Book Is Not

This book is not a statistics textbook. It will reference countless statistical techniques for the data scientists (and interested laymen) among the readers—but it will not explain them. Data scientists know most of the techniques already or at least know where to look them up.

This book is also no legal textbook. It does treat legal and ethical issues on a philosophical level—including how the European General Data Protection Regulation both recognizes and misses some core insights about algorithms—but it does not aim to catalogue all the laws somehow relevant in dealing with algorithmic bias or to give guidance on how to comply with specific legal requirements. This requires lawyers—ideally ones who have read this book, too.

Finally, this book is no silver bullet. Fighting biases is hard. In a sense, biases are a form of conformity—conformity with "the way it is," what your boss says, what the data says, what your lazy mind says (because you always have done it this way). There is zero chance that you will get any benefit from reading this book if you don't change some of the things you are doing. I invite you to keep thinking about what the insights from this book mean to you and what you can do differently because of what you have learned. In fact, I would love to hear it—why don't you leave a comment on my blog at www.linke-din.com/in/tobiasbaer/? And if you want to prevent all your good ideas slipping away and going back to your old ways once you have read this book and found a good space for it on your bookshelf where it can collect plenty of allergenic dust, maybe you even want to put a reminder in your calendar right now!

How This Book Is Organized

The book has four parts. The first part is an introduction—it covers the psychology of human biases as well as how algorithms are used and developed. The chapters of the first part explain the terminology and frameworks I will keep referring to in the remainder of the book.

The second part introduces six distinct sources of algorithms. The understanding of these sources is the basis for managing and preventing algorithmic bias and therefore will be referenced throughout the remainder of the book.

The third part discusses how users of algorithms (broadly defined as anyone who is not a data scientist) can deal with algorithmic bias and what powerful possibilities they have to prevent it.

And the fourth and final part provides comprehensive, practical guidance to data scientists for preventing algorithmic biases through specific techniques for development and implementation. This part of the book is therefore the most technical one—but I still wrote it in a way that everyone from an undergraduate student to a seasoned Head of Analytics can follow and find some valuable insights.

An Introduction to Biases and Algorithms

Introduction

What is a bias? A widely cited source[1] defines it as follows:

> *Inclination or prejudice for or against one person or group, especially in a way considered to be unfair.*

Biases are double-edged swords. As you will see in the next chapter, biases typically are not a character flaw or rare aberration but rather the necessary cost of enabling the human mind to make thousands of decisions every day in a seemingly effortless, ultra-fast manner. Have you ever marveled how you were able to escape a fast-moving object, such as a car about to crash into you, in a split-second? Neuroscientists and psychologists have started to unravel the mysteries of the mind and have found that the brain can achieve this speed only by taking numerous shortcuts.

A shortcut means that the mind will jump to a conclusion (e.g., deem a dish inedible or a stranger dangerous) without giving all facts due consideration. In other words, the mind uses prejudice in order to gain speed.

The use of prejudice in decision-making therefore is unfair insofar as it (willfully) disregards certain facts that may advocate a different decision. For example, if your partner once ate a bouillabaisse fish soup and became terribly sick afterwards, he or she is bound to never eat bouillabaisse again, and may refuse to even try the beautiful bouillabaisse you just cooked, blissfully ignoring the fact that you graduated with distinction from cooking school and bought the best and freshest ingredients available in the country.

---

[1]David Marshall, "Recognizing your unconscious bias," *Business Matters*, www.bmmagazine.co.uk/in-business/recognising-unconscious-bias/, October 22, 2013.

© Tobias Baer 2019
T. Baer, *Understand, Manage, and Prevent Algorithmic Bias*,
https://doi.org/10.1007/978-1-4842-4885-0_1

Algorithms are mathematical equations or other logical rules to solve a specific problem—for example, to decide on a binary question (yes/no) or to estimate an unknown number. Just like the brain making decisions in split-seconds, algorithms promise to give an answer instantaneously (in most cases, the score value of the algorithm's equation can be calculated in a fraction of a second), and they are also a shortcut because they consider only a limited number of factors in a predetermined fashion.

On one level, algorithms are a way for machines to emulate or replace human decision-makers. For example, a bank that needs to approve thousands of loan applications every month may turn to an algorithm applied by a computer instead of human credit officers to underwrite these loans; this often is motivated by an algorithm being both faster and cheaper than a human being.

On another level, however, algorithms also can be a way to reduce or even eliminate bias. Statisticians have developed techniques to develop algorithms specifically under the constraint of being unbiased—for example, the ordinary least squares (OLS) regression is a statistical technique defined as BLUE, the best linear unbiased estimate. Sadly, I had to write that algorithms "*can*" reduce or eliminate bias—algorithms also can be as biased or even worse than human decision-making. Several chapters of this book are dedicated to explaining the many ways an algorithm can be biased.

In the context of algorithms, however, the definition of *bias* should be more specific. Problems solved by algorithms have at least theoretically a correct answer. For example, if I estimate the number of hairs on the head of a well-known president, nobody may ever have counted them, but anyone with unlimited time and access to the president could verify my estimate of 107,817 hairs.

In most situations (including presidential hair), the correct answer cannot be known at least *a priori* (i.e., at the time the algorithm is applied). Algorithms therefore often are a way to make predictions. Through predictions, algorithms help to reduce and to manage uncertainty. For example, if I apply for a loan, the bank doesn't know (yet) whether I will pay back the loan, but if an algorithm tells the bank that the probability of me defaulting on the loan is 5%, the bank can decide whether it will make any profit on me if it gives me the loan at a 5.99% interest rate by comparing the expected loss with the interest charged and other costs incurred by the bank. This illustrates a typical way algorithms are used: algorithms estimate probabilities of specific events (e.g., a customer defaulting on a loan, a car being damaged in an accident, or a person dying by the end of the term of a life insurance contract), and these probabilities allow a business underwriting risks to make an approve/reject decision based on an objective expected risk-adjusted return criterion.

Algorithms are deployed in situations with imperfect information (e.g., the bank's credit rating algorithm doesn't know about the gambling debt I incurred last night, nor does it know if my company will fire me next month). Algorithms therefore *will* make mistakes; however, they are supposed to be correct *on average*. A **bias** is present if the average of all predictions systematically deviates from the correct answer. For example, if the bank's algorithm assigns a 5% probability of default to 10,000 different customers, one would expect that 500 of the 10,000 will default (500/10,000 = 5%). If you investigate the situation and find that in reality 10% of customers default but every time an applicant has a German passport, the algorithm cuts the true estimate by half, the algorithm is biased—in this case, in favor of Germans. (Is it a coincidence that this algorithm was created by a German guy?)

Systematic errors in predictions—whether made by humans or by algorithms— can have serious implications for businesses, and sadly they happen all the time. For example, one study of mega infrastructure projects—analyzing 258 projects in 20 different countries—found cost overruns in almost 9 out of 10 of them, indicative of a systematic underestimation of true cost.[2] During the global financial crisis, banks such as Northern Rock, Lehman Brothers, and Washington Mutual went under because they had systematically underestimated credit, market, and liquidity risks.

Sometimes human bias is to blame. For example, one US bank had an economic capital model (a sophisticated model quantifying those "unexpected losses" of a given portfolio that can cause a bank run or bankruptcy) that prior to the global financial crisis hinted at the out-sized risks looming in home equity loans by estimating unexpected losses many times larger than expected losses; tragically, management dismissed those estimates because they were used to seeing unexpected losses much closer to expected losses and therefore deemed the model to be faulty.

At other times, however, algorithms themselves are flawed. For example, an Asian bank bought a scoring model for consumer credit cards that looked at the card's utilization ratio as one of the predictors of default. The algorithms believed that customers with a low utilization (e.g., using just 10% of the credit limit) were safer than customers with a high utilization; for safe customers, the algorithm increased the limit. However, this created a circular reference: in the moment the algorithm increased the credit limit, the utilization (calculated by dividing the current outstanding balance by the credit limit) dropped, causing the algorithm to further increase the limit (so if the outstanding was 10 and the limit was 100, utilization was 10%; if the system increased the limit by 25% from 100 to 125, utilization dropped to 8% (= 10/125), triggering another increase of the limit, and so on). This happened until credit limits reached stratospheric levels that were totally beyond the customers' means to repay the bank. When more and more customers

---

[2]B. Flyvbjerg, M.S. Holm, and S. Buhl, "Underestimating costs in public works projects: Error or lie?," *Journal of the American Planning Association*, 68(3), 279-295, 2002

started to actually use their very large credit limits, unsurprisingly many defaulted, and the bank almost went bankrupt after having written off more than a billion USD in bad debt.

Algorithmic bias comes in all kinds of shapes and colors. In 2016, ProPublica published a research report showing that COMPAS, an algorithm used by US authorities to estimate the probability of a criminal to re-offend, is racially biased against blacks.[3] MIT reported on natural language processing algorithms being sexist by associating homemakers with women and programmers with men.[4] And research conducted in 2014 showed that setting the user's profile to female in Google's Ad Settings can lead to less high-paying job offers appearing in ads.[5] As more and more decisions are made by algorithms— affecting consumers, companies, employees, governments, the environment, even pets and inanimate objects—the dangers and impact of algorithmic bias is growing day by day. However, this is not by necessity—bias is merely a side-effect of an algorithm's working and therefore a by-product of conscious and unconscious choices made by the creators and users of algorithms. These choices can be revisited and changed in order to reduce or even eliminate algorithmic bias.

This book is about algorithmic bias. First of all, we want to understand better what it is—where it comes from and how it can wreak havoc with important decisions. Second, we want to control its damage by exploring how you can manage algorithmic bias—be it as a user or as a regulator. And third, we want to explore ways for data scientists to prevent algorithmic bias.

The first part, Chapters 2-5, introduces the topic. I will start with a quick review of psychology and human decision biases as algorithmic biases mirror them in more ways than easily meets the eye (Chapter 2) and discuss how algorithms can help to remove such biases from decisions (Chapter 3). Keeping in mind that many readers of this book are laymen and not data scientists, I'll then review how the sausage is made—i.e., how algorithms are developed (Chapter 4) and demystify what is behind machine learning (Chapter 5).

The second part of the book, Chapters 6-11, explores where algorithmic biases come from. Chapter 6 examines how real-world biases can be mirrored by algorithms (rather than rectified). Chapter 7 turns to the persona of the data scientist and how the data scientist's own (human) biases can cause algorithmic biases. Chapter 8 dives deeper into the role of data, and Chapter 9 reviews how the very nature of algorithms introduces so-called stability

---

[3]J. Larson, S. Mattu, L. Kirchner, and J. Angwin, "How we analyzed the COMPAS recidivism algorithm," *ProPublica*, 9, 2016.

[4]W. Knight, "How to Fix Silicon Valley's Sexist Algorithms," *MIT Technology Review*, November 23, 2016.

[5]A. Datta, M.C. Tschantz, and A. Datta, "Automated experiments on ad privacy settings," *Proceedings on Privacy Enhancing Technologies*, 92-112., 2015.

biases. Chapter 10 looks at new biases arising from statistical artifacts, and Chapter 11 deep-dives into social media where human behavior and algorithmic bias can reinforce each other in a particularly diabolical manner.

The third part of the book, Chapters 12-17, approaches algorithmic bias from a user's perspective. It sets out with a brief discussion of whether or not to actually use an algorithm (Chapter 12) and how to assess the severity of the risk of algorithmic bias for a particular decision problem (Chapter 13). Chapter 14 gives an overview of techniques to protect yourself from algorithmic bias. Chapter 15 more specifically describes techniques for diagnosing algorithmic bias, and Chapter 16 discusses managerial strategies for overcoming a bias ingrained in an algorithm (if not real life). Chapter 17 discusses how users of algorithms can make a critical contribution to the debiasing of algorithms by producing unbiased data.

The fourth part of the book, Chapters 18-23, addresses data scientists developing algorithms. Chapter 18 provides an overview of the various ways data scientists can guard against algorithmic bias. Chapter 19 deep-dives into specific techniques to identify biased data. Chapter 20 discusses how to choose between machine learning and other statistical techniques in developing an algorithm in order to minimize algorithmic bias, and Chapter 21 builds on this by proposing hybrid approaches combining the best of both worlds. Chapter 22 discusses how to adapt the debiasing techniques introduced by this book for the case of self-improving machine learning models that require validation "on the fly." And Chapter 23 takes the perspective of a large organization developing numerous algorithms and describes how to embed the best practices for preventing algorithmic bias in a robust model development and deployment process at the institutional level.

Bias in Human Decision-Making

As you will see in the following chapters, algorithmic biases originate in or mirror human cognitive biases in many ways. The best way to start understanding algorithmic biases is therefore to understand human biases. And while colloquially "bias" is often deemed to be a bad thing that considerate, well-meaning people would eschew, it actually is central to the way the human brain works. The reason is that nature needs to solve for three competing objectives simultaneously: accuracy, speed, and (energy) efficiency.

Accuracy is an obvious objective. If you are out hunting for prey but a poorly functioning cognitive system makes you see an animal in every second tree trunk or rock you encounter, you obviously would struggle to hunt down anything edible.

Speed, by contrast, is often overlooked. Survival in the wild often is a matter of milliseconds. If a tiger appears in your field of vision, it takes at least 200 milliseconds until your frontal lobe—the place of logical thinking—recognizes that you are staring at a tiger. At that time, the tiger very well may be leaping at you, and soon after you'll have ended your life as the tiger's breakfast. Our survival as a species may well have hinged on the fact

© Tobias Baer 2019

T. Baer, *Understand, Manage, and Prevent Algorithmic Bias,*
https://doi.org/10.1007/978-1-4842-4885-0_2

that nature managed to bring down the time for the flight-or-fight reflex to kick in to 30-40 milliseconds—a mere 160 milliseconds difference between extinction and by some accounts becoming the crown of the creation! As John Coates describes in great detail in his book *The Hour Between Dog and Wolf*,[1] nature had to go through a mindboggling array of tweaks and tricks to accomplish this. A key aspect of the solution: if in doubt, assume you're seeing a tiger. As you will see, biases are therefore a critical item in nature's toolbox to accelerate decisions.

Efficiency is the least known aspect of nature's approach to thinking and decision-making. Chances are that you grew up believing that logical, conscious thinking is all your brain does. If you only knew! Most thinking is actually done subconsciously. Even what feels like conscious thinking often is a back-and-forth between conscious and subconscious thinking. For example, imagine you want to go out for dinner tonight. Which restaurant would you choose? Please pause here and actually do make a choice! Ready? Have you made your choice? OK. Was it a conscious or subconscious choice? You probably looked at a couple of options and then consciously made a choice. However, how did that short list of options you considered come about? Did you create a spreadsheet to meticulously go through the dozens or thousands of restaurants that exist in your city, assess them based on carefully chosen criteria, and then make a decision? Or did you magically think of a rather short selection of restaurants? That's an example of your subconscious giving a hand to your conscious thinking—it made the job of deciding on a dinner place a lot easier by reducing the choices to a rather short list.

The reason why nature is so obsessed with efficiency is that your logical, conscious thinking is terribly inefficient. The average brain accounts for less than 2% of a person's weight, yet it consumes 20% of the body's energy.[2] That means 20% of the food you obtain and digest goes to powering your brain alone! That's a lot of energy for such a small part of the body. And most of that energy is consumed by the logical thinking you engage in (as opposed to almost effortless subconscious pattern recognition). Just as modern planes and ships have all kinds of technological methods to reduce energy consumption, Mother Nature also embedded all kind of mechanisms into the brain to minimize energy consumption by logical thinking (lest you need to eat 20 steaks per day). Not surprisingly, it introduced all kind of biases through this.

If you collect all the various biases described across the psychological literature, you will find over 100 of them.[3] Many of them are specific realizations of more fundamental principles of how the brain works, however, and therefore several authors have brought down the literature to 4–5 major types of biases.

---

[1]John Coates, *The Hour Between Dog and Wolf*, New York: The Penguin Press, 2012.
[2]Daniel Drubach, *The Brain Explained*. New Jersey: Prentice-Hall, 2000.
[3]Buster Benson, "Cognitive Bias Cheat Sheet," https://betterhumans.coach.me/cognitive-bias-cheat-sheet-55a472476b18, September 1, 2016.

I personally like the framework developed by Dan Lovallo and my former colleague Olivier Sibony:[4] they distinguish action-oriented, stability, pattern-recognition, interest, and social biases. I will loosely follow that framework when in the following I discuss some of the most important biases required for an understanding of algorithmic bias.

Action-Oriented Biases

Action-oriented biases reflect nature's insight that speed is often king. Who do you think is more likely to survive in the wild, the careful planner who will compose a 20-page risk assessment and think through at least five different response options before deciding whether fight or flight would be a better response to the tiger that just appeared five meters in front of him, or the dare-devil that in a split-second decides to fight the tiger?

A couple of biases illustrate the nature of action-oriented biases. To begin with, biases such as the von Restorff effect (focus on the one item that stands out from the other items in front of us) and the bizarreness effect (focus on the item that is most different from what we expect to see) draw our attention to the yellow fur among all those bushes and trees around us; overoptimism and overconfidence then douse the self-doubt that might cause deadly procrastination.

The bizarreness effect can bias our cognition like outliers and leverage points can have an outsized effect in estimating the coefficients of an algorithm. This is because of the availability bias—if we recall one particular data point more easily than other data points (e.g., because it stood out from most other data points), we overestimate the representativeness of the easy-to-remember data point. This can explain why, say, a single incident of a foreigner conducting a spectacular crime can severely bias our perception of people with that foreigner's nationality, causing out-of-proportion hostility and aggression against them.

Overconfidence deserves our special attention because it also goes a long way to explain why not enough is done about biases in general and algorithmic biases in particular. Many researchers have demonstrated overconfidence by asking people how they compare themselves to others.[5] For example, 70% of high school seniors surveyed believed that they have "above average" leadership skills but only 2% believed they were "below average" (where by definition, roughly 50% each should be below and above average, respectively). On

---

[4]D. Lovallo and O. Sibony, "The case for behavioral strategy," *McKinsey Quarterly*, 2(1), 30-43, 2010.
[5]The examples here are taken from and further referenced in D. Dunning, C. Heath, and J.M. Suls, "Flawed self-assessment: Implications for health, education, and the workplace," *Psychological science in the public interest*, 5(3), 69-106, 2010.

their ability to get along with others, 60% even believed to be in the top 10% and 25% in the top 1%. Similar results have been found for technical skills such as driving and software programming. Overoptimism is essentially the same bias but applied to the assessment of outcomes and events, such as whether a large construction project will be able to remain within its cost budget.

What does this mean for fighting bias? Even if people accept the fact that others may be biased, they overestimate their own ability to withstand biases when judging—and as a result resist efforts to debias their own decisions. With most people succumbing to overoptimism, we can easily have a situation where most people accept that biases exist but still the majority refuses to do anything about it.

Another fascinating aspect of the research of overoptimism: it has been found in Western culture but not in the Far East.[6] This illustrates that both individual personality and the overall culture of a country (or company/organization) will have an impact on the way we make decisions and thus on biases. A bias we observe in one context may not occur in another—but other biases might arise instead.

■ **Note** An excellent demonstration of overconfidence is the fact that I observe that because of overconfidence, most people fail to take action to debias their decisions—but I write a book on debiasing algorithms anyhow, somehow believing that against all odds I will be able to overcome human bias among my readers and compel them to implement my suggestions. However, I also know that *you*, my dear reader, *are* different from the average reader and a lot more prone to actually take actions than others; therefore, let me just point out that in order to be consistent with your well-deserved positive self-image, you should make an action plan today of how you will apply the insights and recommendations from this book in your daily work and actively resist the tempting belief that you are immune to bias, lest you fail to meet the high expectations of both of us in our own respective skills.☺

Stability Biases

Stability biases are a way for nature to be efficient. Imagine you find yourself the sole visitor at the matinee showing of an art movie—you therefore could choose literally any of the 200 seats. What would you do: jump up every 30 seconds to try out a different one, or pretty much settle into one seat, at

---

[6]It is a general limitation of social psychology that most empirical research is done within the context of Western culture, with a substantial portion of the research carried out even more narrowly with North American college students. The few studies testing Western theories in Asian cultures such as Japan or China regularly find important cultural differences.

most changing it once or twice to maybe gain more legroom or escape the cold breeze of an obnoxious air conditioning? From nature's perspective, every time you just think about changing your seat, you have already burned mental fuel, and if you actually get up to change a seat, your muscles consume costly energy, let alone that you might be missing the best scene of the movie. A number of biases try to prevent waste of mental and physical resources by "gluing" you to the status quo.

Examples for these biases include the status quo bias and loss aversion. You like the seat you are sitting on better than other seats simply because it is the status quo—and you hate the idea of losing it. This is a specific manifestation of loss aversion that is dubbed the *endowment effect*; it has been shown in experiments involving university coffee mugs and pens that once an object is in your possession (i.e., you are "endowed" with the object), the minimum price at which you are willing to sell might be roughly *double* the maximum price you would be willing to pay for the item.[7]

While economists consider such a situation irrational and abnormal, from nature's perspective it appears perfectly reasonable—nature wants you to either take a rest or do more productive things than trading petty items at negligible personal gain! At times, however, this status quo bias overshoots. For example, corporate decisions in annual budgeting exhibit a very strong status quo bias, with one analysis reporting a 90% correlation in budget allocations year after year (of individual departments or units). While this might have avoided an acrimonious debate of taking away budget from some units, this stability comes at enormous economic cost: companies with more dynamic budget allocation grow twice as fast as those ceding to the status quos bias.[8]

Another important stability bias is the anchoring effect. Econometricians studying time series models often are surprised at how well the so-called naïve model works[9]—for many time series, this period's value is an excellent predictor of the next period's value, and many complex time series models barely outperform this naïve model. Nature must have taken notice because when humans make an estimate, they often root it heavily in whatever initial value they have and make only minor adjustments if new information arises over time. At times, this bias leads seriously astray, however—namely if the initial value is seriously wrong or simply random. A popular demonstration of the anchoring effect involves asking participants to write down the last two digits of their social security or telephone number before estimating the price

---

[7]D. Kahneman, J.L Knetsch, and R.H. Thaler, "Anomalies: The endowment effect, loss aversion, and status quo bias," *Journal of Economic Perspectives*, 5(1), 193-206, 1991.

[8]T. Baer, S. Heiligtag, and H. Samandari, *The business logic in debiasing*, McKinsey & Co, 2017.

[9]https://blogs.sas.com/content/forecasting/2014/04/30/a-naive-forecast-is-not-necessarily-bad/

of an item, such as a bottle of wine or a box of chocolates. Even though there is obviously absolutely no relationship with these numbers and the price of the item, those writing down high numbers consistently estimate prices 60 to 120 percent higher than those with low numbers.[10]

Pattern-Recognition Biases

Pattern-recognition biases deal with a very vexing problem for our recognition: much of our sensual perception is incomplete, and there is a lot of noise in what we perceive. Imagine the last time you talked with someone—probably it was just a few minutes ago, maybe you spoke to the train conductor or the flight attendant if you're reading this book on the go. Think of a meaty, information-rich sentence the other person said in the middle of the conversation. Very possibly a part of the sentence was actually completely drowned out by a loud noise (e.g., another person's sneeze), several syllables might have been mumbled, and you also may have missed part of the sentence because you glanced at your phone. Did you ask the person to repeat the sentence? Or did you somehow still have a good idea of what the person said? Very often it's the latter—because of an amazing ability of our brain to "fill in the gaps." Our brains excel at very educated guessing—but sometimes these guesses are systematically wrong, and this is the realm of pattern-recognition biases.

Pattern-recognition biases are particularly relevant to this book because pattern-recognition is essentially what algorithms do.

In order to solve the problem of making sense from noisy, incomplete data (be it visual or other sensual perception, or be it actual data such as a management information system report full of tables in small print), the brain needs to develop rules. Systematic errors (i.e., biases) occur if either the rules are wrong or a rule is wrongly applied.

The Texas Sharpshooter fallacy is an example of a flawed rule. Your brain sees rules (i.e., patterns) in the data where none exists. This might explain many superstitions. If for three times in a row a sales person closes a deal while wearing the red tie she got from her husband for her birthday, the brain might jump to a conclusion that it is a "lucky tie." Interestingly, the brain may not be wrong—it's possible that the color red has a psychological effect on buyers that does increase the odds of closing the deal—it's just that three closed deals is a statistically insignificant sample and way too little data to make any robust inference. This illustrates that the way nature thinks about pattern recognition is heavily driven by a "rather safe than sorry" mentality—how many times does the neighbor's dog have to bite you in order for you to conclude that you better not get anywhere close to this cute pooch? By the

---

[10]E. Teach, "Avoiding Decision Traps," *CFO*, June 1, 2004; Retrieved October 29, 2018.

same token, the brain is hardwired to think that even if there is only a small chance that the red tie helps, why risk a big deal by not wearing it?

Confirmation bias can be an accomplice of the Texas Sharpshooter fallacy and is nature's way of being efficient in the recognition of patterns. Confirmation bias can be seen as a "hypothesis driven" approach to collecting data points. It means that where the mind has a hypothesis (e.g., you already have a belief that buying this book was a great idea), you tend to single out new data that confirms your belief (e.g., you praise the five-star review of this book for its brilliant insights) and reject contradictory data (e.g., you label the author of the one-star review a fool—of course rightly so, may I hasten to say!). Underneath the confirmation bias seems to be nature's desire to get to a decision quickly and to reduce cognitive effort. Laboratory experiments have shown that participants are much more likely to read news articles that support their views than contradictory ones. You'll therefore encounter confirmation bias as a central foe in Chapter 11 about algorithmic bias in social media.

Confirmation bias also can shape how we process "noisy" information. Imagine the above mentioned interaction with a flight attendant or train conductor. She asked about the book you were reading and you proudly showed her the cover of this book. Just as she replied, a loud noise drowned out part of her sentence. There really is no way to tell if she said "I loved that book!" or "I loathed that book!" Except that most likely you "heard" her say that she loved the book. This is because your brain of course would have expected her to say so, and an inconclusive sound would be automatically and subconsciously replaced with the expected content.

Stereotyping is an extension of the confirmation bias and an example of a bias where a rule is applied overly rigidly. First, imagine that you are in a swanky restaurant. The waiter just brought the check to the table next to you where now a stately, senior, white man pulls out a black object from his pants. What do you think it is? You probably imagined a wallet. Now imagine a police car passing a visibly distressed woman lying on the side of the street. As the police car pulls by, the woman shouts "my purse, my purse!" and waves into the air. At this moment, the police officers become aware of a young black man nearby running towards a subway station. They immediately run after the man, shouting "Stop! Police!" and aim their guns at the man. As the man reaches the steps of the entrance of the subway station, he pulls a black object out of his pocket. What is it? If you imagined a gun (not the wallet containing the man's subway pass, which he needs to produce quickly if he doesn't want to miss his train and hence arrive late for his piano lesson), then you fell victim to stereotyping. Based on the situation's context, your brain already has some expectations of what reasonably could happen next. A person in a restaurant who just received a check is likely to pull out a wallet, credit card, or bundle of banknotes from his pocket; a person who appears to have committed a robbery is likely to pull out a knife, gun, or hand grenade from his pocket when trying to escape from the police. When all the brain knows is that a "black object" is pulled from the pocket, it "fills in the gaps" based on these stereotyped views of what a person in such a context is most likely to have in his pocket. The dilemma is that this guess might be wrong. It is quite clear that a police officer who shoots a suspect in the moment a black object is drawn from the pocket is less likely to be shot and hence more likely to survive than a more careful and deliberate officer who doesn't pull the trigger unless the suspect has without any doubt pointed a gun at him, so evolution has not exactly been the greatest fan of due process. However, if the officer ends up shooting an innocent because the innocent looks like the "stereotypical" robber in the officer's mind, nature's trickery and decision bias have tragically claimed a life.

This example also points at the sober fact that sometimes the creation and use of algorithms can require us to decide grave ethical dilemmas—be it the denial of human rights to suspected terrorists, the decision to give parole to a convicted criminal, or the programming of self-driving cars for a situation where a deadly collision with pedestrians is unavoidable but the car could decide which of several pedestrians to run over. Just like in a classic tragedy where the hero had to choose between two equally disastrous paths, algorithms sometimes must be predisposed to go in one or the other direction, and the "bias" we ultimately decide to embed in the algorithm reflects our best assessment of what the most ethical decision is in this case.

Interest Biases

Interest biases go beyond mere heuristic shortcuts. Where action, stability, and pattern-recognition biases simply aim at making the "correct" decision as accurately, quickly, and efficiently as possible, interest biases explicitly consider the question "What do I want?" Think for a moment of a so-so restaurant in your area where you would go only reluctantly. Now imagine that a friend asked you to go there for lunch tomorrow. What thoughts came to your mind? Were you quick to think "rather not—what about we go to …?" Now imagine that your credit card company has made a fantastic offer that if you charge a meal for two at this so-so restaurant on your card by the end of the week, you get a $500 shopping voucher. Now imagine again that your friend asks you out for lunch—however, this time it is to a different restaurant that in general you like a lot. What is your reaction now? Can you suddenly sense a desire to go to the so-so restaurant instead?

If you listen very carefully to your thoughts, you might realize that your subconscious influences your thought process in quite subtle ways. In our exercise, you may have thought not just how you would suggest to your friend to go to a different restaurant but your subconscious also might have supplied several talking points to buttress your suggestion (i.e., additional talking points against or in favor of the so-so restaurant). If I task you with not liking a given restaurant, your mind is bound to retrieve in particular those attributes of the restaurant that you don't like. By contrast, if I task you to sense a desire to go to the same restaurant, your mind is bound to retrieve the more attractive attributes of the restaurant.

The point is that not only might you rationally prefer the so-so restaurant if going there promises a windfall gain of $500 but that this explicit or hidden preference influences your assessment more broadly (think of it as a confirmation bias) and you might therefore seriously believe that the restaurant was a good choice also for your friend (who sadly won't get a shopping voucher). Interest biases therefore can considerably muddy the water—rather than objectively stating that an option would be good for you but bad for everyone else, your own mind might wrongly convince you that your preferred option is objectively superior for every single stakeholder.

One can observe such behavior frequently in corporate settings where certain decisions have very personal implications. Imagine that your company is considering moving its offices to the opposite side of the town. How do you feel about it? Do you think there is any correlation between your overall assessment of the move and whether you personally would prefer the office to be on this or the other side of town? Interest biases therefore greatly influence the behavior of data scientists as well as users of algorithms.

Social Biases

Social biases are arguably just a subcategory of interest biases—but they are so important that I believe that it is justified to call them out as a separate group. Humans are social beings and, except for a handful of successful hermits, unable to survive outside of their group. Put bluntly, if a caveman annoyed his fellow cavemen so badly that they expelled him from the cave, he soon would be eaten by a wild animal. Humans therefore have an essential fear of ostracism that in essence is a fear of death—in fact, social exclusion causes pain (as a warning signal of imminent harm) that is stronger than most forms of physical pain (which also explains why increasing loneliness is a public health crisis and the UK appointed a "minister of loneliness.")

In decision-making, the mind therefore weighs the benefits of any possible action against the risk of being ostracized. *Sunflower management* is the bias to agree with one's boss; *groupthink* is the bias to agree with the consensus of the group. Members of committees tasked with important decisions frequently admit that they have supported a decision they personally deemed gravely wrong, maybe even disastrous (think of M&A decisions that destroyed otherwise healthy companies) because they deemed it socially unacceptable to veto it.

Similar to other interest biases, social biases work on two levels. While frequently enough people know perfectly well what the correct answer is and still go with an alternative decision because they believe that it would be suicidal to speak the truth, social biases also affect cognitive processes more broadly—for example, they may trigger a confirmation bias where committees strive to find numerous pieces of evidence supporting the chairman's views while subconsciously rejecting evidence that he may be wrong.

Specific Decisions vs. the Big Picture

Interest and social biases illustrate that nature often looks at a bigger picture and not just a single decision. This observation has also been used as a criticism against the very idea of cognitive bias—the argument goes that the concept of a bias is an illusion because we just look at one particular outcome from one particular decision (e.g., if in a particular experiment the participant correctly estimated the number we asked for) whereas nature looks at a much bigger picture in the context of which an individual's behavior and decision-making is perfectly rational and correct.

We don't need to get lost in debating the merit of every single cognitive effect described as a bias by the psychological literature; what matters is the recognition that "truth" (or in the science of statistical modeling, a binary outcome of 0 or 1, the definition of 1) often is a surprisingly context-driven concept and that where a decision appears biased it simply may be the reflection of a compromise between competing objectives—be it simply a consideration of the importance of speed and efficiency, or a recognition that a single decision might be a negligible element in a much larger and complex problem such as maintaining good social relationships.

This realization is important for the detection, management, and avoidance of bias. As you will see in subsequent chapters, not every time that a predictive outcome (and by implication the responsible underlying algorithm) appears biased does this constitute a "problem" that needs to be fixed. At the same time, behaviors leading to (wrongly) biased outcomes still might be justified or at least rational, and therefore an effective way to contain such a bias will have to work around such behavior rather than expecting the behavior itself to change. And whenever a data scientist takes an action to explicitly eliminate a particular bias from an algorithm, he or she may have to assess whether doing so may cause a different problem in the system at large, and if therefore the intervention is warranted and really in the best interest of the stakeholders supposedly benefiting from debiasing the algorithm.

Summary

In this chapter, you saw how biases originated in a need for speed and efficiency in decision-making or other personal interests including the need for social inclusion that sometimes trumps the need for accuracy. These biases shape many phenomena that algorithms describe and predict; they also shape the behavior of humans that create, use, or regulate algorithms. As you will see in subsequent chapters, the techniques used to develop algorithms can eliminate some of these biases while other biases are bound to be mirrored by algorithms. Finally, this overview of human cognitive biases will also be useful in understanding other types of biases that are specific to algorithms due to the way they are developed and deployed.

The most important biases for our purposes are:

- *Action-oriented biases* are driving us to speedy action by focusing attention and deflecting procrastination due to self-doubt.

- An *availability bias* is a particular action-oriented bias that lets specific data points—especially bizarre or otherwise noteworthy ones—inordinately shape our predictions.

- *Overoptimism* and *overconfidence* are specific action-oriented biases that can cause developers and users of algorithms to disregard dangers and limitations and to overestimate the predictive power of their algorithms.

- *Stability biases* minimize cognitive and physical efforts by gluing us to the status quo.

- *Anchoring* is a specific stability bias that can compromise estimates by rooting them in completely random or seriously flawed reference points.

- *Pattern-recognition biases* lead to flawed predictions by either forming decision rules from random patterns or by applying sensible rules inappropriately.

- *Confirmation bias* is a particular pattern-recognition bias that compromises the data we consider when developing pattern-based decision rules.

- *Interest biases* compromise objective judgment by blending it with our self-interest.

- *Social biases* are a particular type of interest biases that focus specifically on safeguarding our standing within our social environment.

We now turn to algorithms and start unraveling the many ways they are affected by and interact with these biases. As a first step, we need to disentangle statistical algorithms in general from a particular type of algorithm, namely those developed through machine learning.

How Algorithms Debias Decisions

In the previous chapter, you took a crash course in psychology to understand why humans are sometimes biased in their decision-making and what some of the most prevalent biases are. In this chapter, aimed primarily at readers who do not have experience in building algorithms themselves, I will explain how an algorithm works. More specifically, I will show how a *good* algorithm works and how it thereby can *alleviate* human bias; in later chapters you then can understand more easily how algorithms can go awry (i.e., show bias)—and what the different options are for addressing this problem.

A Simple Example of an Algorithm

In the introduction, I mentioned that algorithms are statistical formulas that aim to make *unbiased* decisions. How exactly do they manage to do this?

T. Baer, *Understand, Manage, and Prevent Algorithmic Bias*,
https://doi.org/10.1007/978-1-4842-4885-0_3

One of the simplest statistical algorithms is the linear regression. You estimate a number—for example, the number of hairs on a person's head—with an equation that looks like this:

$$y = c + \beta_1 \cdot x_1 + \beta_2 \cdot x_2 + \beta_3 \cdot x_3$$

The *dependent* variable (i.e., the number you want to estimate, here the number of hairs on a person's head) is often denoted as y. It is estimated as a linear combination of *independent* variables (also called *predictors*). Here the data scientist has chosen three predictors: x_1, x_2, and x_3. x_1 might be the surface area of the scalp of the person (larger heads have more hair), x_2 might be the age of the person (as people get older, they may lose some hair), and x_3 may be the gender (men seem to be bald more often than women).

How does this equation work? Let's assume that this morning you marked one square centimeter on your Mom's head and counted 281 hairs. You therefore may start off by multiplying the surface area of the scalp (denoted as x_1 and measured in square centimeters) by 281. Statisticians call 281 a *coefficient* and denote it as β_1. The subscript of 1 simply indicates that β_1 belongs to x_1.

Extrapolating from the number of hairs you found this morning in your bed, you also may believe that every year people lose on average 1,000 hairs. You therefore put −1,000 into β_2.

Finally, you guess that men on average have 50,000 hairs less than women. But here you run into a problem: gender is a qualitative attribute, but your equation needs numbers. How can you solve that? The solution is what data scientists call *variable transformation* or *feature generation*: they create new numerical values that measure qualitative attributes (or other things—transforming qualitative values into numbers is just an example of feature generation). In our example, they would define x_3 as a numerical value indicating "maleness." The simplest way is a dummy variable: x_3 can be defined as a binary variable that is 1 for men and 0 otherwise. In this case, β_3 could be set to −50,000.

A different data scientist, however, might argue that a binary definition of gender is outdated and too crude and therefore suggest that "maleness" is measured by the testosterone level in the blood. In this case, the model might suggest that for every 1 ng/dL testosterone the number of hairs decreases by 70. Note that you have encountered the first example of how the beliefs of the data scientist—here on whether gender is binary or not—can shape an algorithm.

By looking at a very limited amount of data—one square centimeter of your Mom's head, the hair found in your bed this morning, and some estimate on gender differences you pretty much pulled out of thin air—you therefore have come up with an algorithm:

$$\text{hair} = 281 \cdot x_1 - 1,000 \cdot x_2 - 70 \cdot x_3$$

The problem with this algorithm is that it is pretty wrong. Your Mom might be rather exceptional; your count might have been flawed; and conceptually, you ignored the fact that men have larger heads than women, so there is what statisticians call some correlation between x_1 and x_3 (and when you chose the coefficient for x_3, you didn't think about what x_1 does at the other end of your equation). Equipped with the knowledge of the previous chapter, you can see how your cognitive biases might have helped to create this mess: you fell victim to the *availability bias* by basing your data collection on your immediate family members and exhibited extreme *anchoring* by not even considering your neighbor. And you exhibited gross *overconfidence* by believing this approach actually made any sense!

Luckily for you, statistics can come to the rescue: if you measure the amount of hairs as well as x_1, x_2, and x_3 for a sample of people (your statistician friend might suggest at least 100-200 people, although a million people would be much better), statistics will optimize the values of your coefficients in such a way that the estimation error is minimized. The statistical estimation procedure will "play around" with and find the optimal estimates for four parameters: β_1, β_1, and β_3 as well as the constant parameter c.

Let me make a technical note here: linear regression is still so simple statistically that using matrix algebra, the coefficients actually can be calculated. For more complex algorithms, however, "playing around" is indeed the only way to find a good solution. Techniques such as a maximum likelihood estimation find optimum solutions for the parameters iteratively. This hints at why the proliferation of advanced statistical techniques depended so much on the ever increasing speed of computers—and why one element of a data scientist's skill (or the skill of the software tool she uses) lies in the computational efficiency, such as knowing clever ways for the maximum likelihood estimation to speed up the search process.

One of the greatest things about algorithms developed by statistics is that they speak to us—sort of. By reviewing the statistical outputs, we can learn not only a lot about the data and the phenomena we try to predict but also how the algorithm "thinks." I will illustrate this briefly in the next sections.

What an Algorithm Will Give You

If you were to collect data on the hair of 200 people (I doubted my publisher would foot the bill for that, so I had to completely make up the following numbers), your statistical software might produce the following equation:

$$hair = 75{,}347 + 159 \cdot x_1 - 0.3 \cdot x_2 - 23 \cdot x_3$$

This is an exceedingly interesting result! There are at least three particularly noteworthy aspects of this equation:

1.   This new equation makes a lot less errors than the original equation (you don't see that but take my word for it). In fact, the estimation technique guarantees that it is the best estimate you can get (remember the concept of BLUE (best linear unbiased estimate) from Chapter 1?). If your livelihood depends on knowing people's number of hairs (e.g., because you are a mad barber charging by the number of hairs cut but your customers reasonably require a quote before you get started), you will be much better off—very possibly you could not even run your business without this equation. However, there is a caveat: "error" is defined in one particular way, namely the squared difference between the algorithm's estimate of the number of hairs and the actual number of hairs of each person in the sample (this is why statisticians also call this technique as ordinary least squares or OLS). That means that if you think differently about weighing errors (because OLS squares the errors, it penalizes large errors but discounts small errors—you might disagree with this), you might prefer a different set of coefficients.

2.   You may have noticed that there is a very large number in the constant term while β_1 is much lower than in the original equation. Essentially the algorithm grounds itself in an anchor (getting it half way to the average) and limits the amount of variation caused by individual attributes. This smacks of the stability bias and echoes the discussion in the previous section where I observed that nature had a point when it hardwired stability biases in our brains—some amount of anchoring does improve the estimates. Empirically we can observe that the poorer the predictors or the overall structure of the model, the more the equation will ground itself in the population average (in the extreme case where the independent variables are all meaningless, the equation will be a constant function with zero coefficients for all independent variables and thus estimate the population average for everyone, not a bad idea under such circumstances).

3.   Finally, you might be struck by the coefficient of −0.3 for age (yep, that's not even half a hair!). If a statistician examined this estimate more closely, she might tell you that it is "insignificant" (i.e., that most likely the coefficient

is zero). A zero coefficient is how statistics tells you that a variable doesn't make sense. Upon reflection, you can see why: in early age, you grow more hair, not less; and even as an adult, you not only lose hair (which then can be found in your bed in the morning) but also grow new hair. Most people therefore encounter a net hair loss only rather late in their life. The linear model simply assumes that every year you lose the same amount of hair—and therefore it really doesn't make much sense. And it is important to understand that because of small sample sizes, estimated coefficients barely are *exactly* zero; coefficients usually contain a little bit of noise. The concept of significance unfortunately is a little bit complicated but what a statistician means by saying that a coefficient is insignificant is the following: if in reality the coefficient was zero (i.e., the variable was meaningless), because of the noise in the data and the sample size you have, you should expect the calculated coefficient to be in a range from here to there (in this example, maybe –0.5 to +0.5). As the actual coefficient you got happens to be within this range, you should assume that in reality it is zero, and therefore that the variable is meaningless. Is this certain? No, it isn't! However, statisticians have tools to quantify the probability of their statements being correct—and in this context, we call this probability the *confidence level*. If a variable is said to be insignificant at the 99.9% confidence, it means that if the true coefficient is zero, estimates have a 99.9% probability to be within a specific range (which is calculated by the statistical software), and the coefficient you got for the variable actually is within that range. In fact, you also could ask the statistician, "If the true coefficient was -1,000, what is the probability that I would get the result we have here?" The statistician might look puzzled because this is not the way people usually think about it—but after some grumbling, he might tell you that it's 0.0000417%. This means that it's really unlikely that we lose in average 1,000 hairs per year—but based on your sample of 200 people, it's not entirely impossible.

The last thing mentioned—telling us that a variable most likely is meaningless—is one of the most important ways a statistical algorithm helps us to debias our decision logic. Statisticians often say that statistical testing "rejects" our hypothesis: it teaches us that based on the data at hand, our belief appears to be wrong.

I spent many years as a consultant debiasing judgmental credit and insurance underwriting. In order to do this, I spent many hours with underwriters to list out all the information they look at; typically we arrived at long lists with 200-400 factors. I then worked with them to prioritize and short-list the 40-70 factors they believed to be most important. And then I applied statistics to validate these factors—and every single time, in over 100 such studies, about half of the prioritized factors turned out to be insignificant. Underwriters suffered from the whole list of cognitive biases I listed in the previous chapter. For example, if a Chinese credit officer once had a German borrower who defaulted in a spectacular bankruptcy, the bizarreness effect might cause this credit officer to reject all German applicants as exceedingly risky. As a German, I can assure you that this would be insane!

Based on the statistical test results, I then redesigned the assessment logic. While judgmental risk assessment typically happens in an underwriter's head, I replaced it with a statistical algorithm. When the banks and insurance companies I worked for tested the new algorithms, they found that decisions were consistently better to the tune of reducing credit and insurance losses by 30-50% (sometimes more) while *at the same time* approving more deals (and hence enjoying faster growth of their business)! This is obviously worth a lot of money to financial institutions (and made paying for my services a really great investment).

For this reason, it is fair to say that statistical algorithms are an important tool in fighting biases. As you will see in subsequent chapters, however, this sadly does *not* mean that algorithms are perfect or cannot fall victim to biases themselves.

Summary

In this chapter, you explored what an algorithm is and identified a few important properties as far as biases are concerned:

- In principle, statistical algorithms aim to make unbiased predictions; they do this by objectively analyzing all data points given to them.

- Inspecting algorithms can offer important information—we not only learn more about the data and the phenomena we want to predict but we can also understand how the algorithm "thinks." This is valuable because it allows us to then ask if that thinking might be flawed.

- Where we enter specific variables in a model (as manifestations of beliefs on underlying causal relationships), statistical *significance* allows us to test whether our hypotheses are supported by the data. If not, chances are that we were biased and the algorithm helped us to detect this bias.

- Empirically, statistical algorithms often perform better than subjective human judgment because they succeed in eliminating many biases.

- Statistical algorithms are anchored in the population average, and the worse the model structure and the predictive power of the independent variables, the stronger this stability bias is.

In order to understand how biases can still sneak into a statistical algorithm, it helps to know in greater detail how data scientists actually go about developing these algorithms. This is what we will consider next.

The Model Development Process

In the previous chapter, you saw how an algorithm works. In this chapter, I will review how an algorithm is developed; this obviously is hugely helpful in understanding the many ways biases can creep into algorithms. Also, seasoned data scientists may want to briefly glance at this chapter so that they are aware of my mental frame and terminology since I will be referencing both frequently going forward. One note on terminology: with the advent of machine learning, a whole new vocabulary has been introduced (e.g., *observations* have become *instances*, *dependent variables* have become *labels*, and *predictive variables* have become *features*), which unfortunately makes it really hard to write something that all generations of data scientists can understand. At least the new job title of *data scientist* is a lot fancier than *model developer* or *modeler*, which is what data scientists used to be called in ancient times (ca. anno 2010)! Apart from the title, I will generally use more traditional terms, mostly for the benefit of those who may have had just a tiny bit of exposure to statistics in other fields of study and for whom it will be easier to connect the dots if I use familiar terms.

© Tobias Baer 2019
T. Baer, *Understand, Manage, and Prevent Algorithmic Bias*,
https://doi.org/10.1007/978-1-4842-4885-0_4

You may wonder why the title of this chapter talks about developing a *model* as opposed to an *algorithm*. When data scientists estimate the parameters of an equation to predict a specific outcome in a particular context (e.g., predict the probability of default for a bank's Canadian small business retail portfolio), we typically call the outcome a *model*. Terminology often varies a bit by industry, geography, function (e.g., risk management or marketing), and even organization; some readers (e.g., folks working in banking) will have heard the term *model* (e.g., US banking regulators use this term to regulate banks' algorithms) while others may be used to the term *algorithm* (e.g., folks dealing with specific functions on a website such as a recommendation engine). For our purposes, there is no need to differentiate between the terms *model* and *algorithm*.

Overview of the Model Development Process

On a high level, one can differentiate five major steps in model development: model design, data engineering, model assembly, model validation, and model implementation.

1. *Model design* defines the overall structure of the model, such as what shall go in and what shall come out of it— not unlike the plan an architect makes for the construction of a new house.

2. *Data engineering* prepares the data used to estimate the coefficients of the algorithm. It covers all activities from identifying the data you want to collect (in our architecture analogy, this first substep is placing an order for construction materials) to putting all data neatly into one or more large tables—with most challenges hiding behind the notion of "neatly." (Just think of bathroom tiles—if you want to have that perfect bathroom, the tiler should carefully inspect each tile, dispose of the broken ones, and cut some tiles just to the right length to fit into corners and crevices.)

3. *Model assembly* is the heart of model development. Here the raw data is transformed into an equation, with coefficients derived through statistical techniques.

4. *Model validation* is an independent review and assertion of the model's fitness for use.

5. *Model implementation* is the deployment of the model in actual business operations.

Let's discuss each step in a bit more detail, in particular the two steps most important for taming biases: data engineering and model assembly.

Step 1: Model Design

Model design determines the big, fundamental questions about a model. We can call them the "four whats:"

- What outcome is predicted...
- for what kind of business problems...
- based on what kind of data...
- with what kind of methodology?

The answers to these questions are really driven by the needs of the business users and the way they intend to use the model; just as an architect may design a useless building if there is insufficient understanding of the customer's needs, a lot of biases can creep into a model if there is insufficient communication between the data scientist and the business users.

Step 2: Data Engineering

Just as some of the best Parisian chefs pride themselves on shopping the Marché International de Rungis at some ungodly hour to get the best pick of the very best local produce (the market opens at 1 a.m. and closes at 11 a.m.), much of the value created by the data scientist happens in the data engineering phase. While different data scientists use slightly different terminology and groupings for their activities, I find it most useful to distinguish five major elements of data engineering: sample definition, data collection, splitting of samples, (addressing of) data quality, and data aggregation.

- *Sample definition* determines exactly which historical reference cases to collect data on. In your hair example, you picked 200 reference cases—should they all be in your neighborhood or shall you include some data points from a different city or even different country? Should you make sure to include Caucasians, Blacks, and Latinos in certain ratios—a process called *stratification*? What about Pacific Islanders? Should you also stratify by age and gender? And would it be better to collect data on 500, 50,000 or maybe 5 million people? Sample definition quickly gets complex, and any trade-off you make can quickly come back to bite you! Only sales people of some overpriced software tool will claim that this is easy—in my experience, many of the gravest problems of algorithms are rooted in poor sampling.

- *Data collection* is the process of getting the actual data for your sample. In the old days, it often involved the IT department writing queries in COBOL for mainframe computers or retrieving data tapes from the dusty archive in the basement; nowadays the data scientist may just type up a simple query into the data lake. Data may have to be collated from multiple sources; sometimes it even may have to be captured manually (e.g., looked up in paper files and typed into a spreadsheet).

- *Splitting of the sample into development, test, and validation samples* is critical to allow a proper validation of the model (a key technique to ensure that the model functions properly). You are in trouble if you forget this or make a poor choice in the way you split (e.g., leaving you with too little or inappropriate data for validation). The model coefficients are estimated based on the development sample; if it fails to show a similar predictive power on the test sample, the data scientist knows that the model is overfitted to the development sample and hence unstable, and can adjust the model. If there are many iterations, however, the model may become overfitted to the test sample as well, and therefore a separate validation sample that is not touched before the final validation is the ultimate test of the model's stability. A paranoid user may not share the development sample with the data scientist before the model is finished (this is how modeling competitions are run). By contrast, if your data scientist forgets to split the sample at the very beginning, the integrity of the thought process is ruined even if later a validation sample is set aside. This is a bit like reading your child's diary: once you have read it, you have broken her trust, even if you put back the diary on the shelf *exactly* how it was.

- *Data quality* needs to be first assessed and, once the specific issues of the data collected have been identified (e.g., missing data or non-sensical values), achieved through a detested activity called *data cleaning*. For example, if you look at a sample of individuals on November 1, 2018 and find that half of them are exactly 118 years and 10 months old, this is suspicious—and closer investigation might find that for many people in the database the date of birth was not collected but some really old computer system three mergers ago entered January 1, 1900 as a default date of birth in such cases. You then either need to overwrite every incidence of January 1, 1900 with a "missing" indicator, or if you are really lucky, you realize that these are all Chinese citizens for which you also have their ID number where the date of birth happens to be encoded—so you can find out the correct age for each of them and thus *clean up* this data quality issue by overwriting January 1, 1900 with the correct date of birth taken from the ID number. Why is this activity detested, though? In doing your clean-up, you might stumble on the fact that for many other people (with more reasonable dates of birth) the ID shows a different date of birth as well. Annoyed, you might decide to replace the date of birth for *all* people based on their ID. At this point, however, you find several people born in 2058, which at the time of writing of this book is still in the future. An investigation of their ID numbers reveals that the check digit of the ID number is wrong and hence the ID must have been incorrectly captured. You sigh and realize that this will never end and that you *really* hate data cleaning.

- *Data aggregation* combines multiple data items (e.g., individual transactions you have made with a credit card or items in your browser's search history) into new variables. This is a key step that at the same time can create more insightful variables (e.g., it might be more meaningful to know that on average you spend $1,287 per month on food than knowing that yesterday you spent $1.99 at Wholefood Markets—forgot the whipping cream, huh?) and lose information (the fact that in three out of four cases you will return to the same supermarket within five hours after a major purchase with a transaction value above $50 to make another small purchase worth less than $10 is actually a very important insight that may suggest to a bank that you are a lot more likely to forget paying your credit card than the average customer—so if I aggregate all food purchases to a total amount for each day or month, I lose the information that you're a real scatterbrain).

In most model development efforts, data engineering is the most time-consuming work step; as you will see soon, it's also ripe with opportunity to create bias.

Step 3: Model Assembly

Once the data is prepared, data scientists can start to assemble the algorithm. This entails a lot more than just running a statistical software package to estimate coefficients of an equation—in fact, it entails seven substeps. Steps 2, 4, and 5 are the fun part that data scientists typically enjoy the most; it therefore may be no coincidence that they sometimes get carried away with these steps and then run out of time for some of the other steps.

I also should note that *model assembly* is not a very common term; usually you will hear *model estimation*—but my whole point here is that model estimation is just one of seven important steps, and biases often creep into models if the other steps are forgotten or shortchanged.

- *Exclusion of records* based on logical criteria is a key step to prevent bias. Many samples contain hidden garbage that slips through if data scientists don't spend enough time on this step. For example, you naively might assume that to build a credit risk model, you simply look at past loans, classify them as either repaid or defaulted, and build a model. Big mistake! Because of rounding issues, many banks have loan accounts in their books that the customer believed to have completely repaid but that technically still have a cent or so outstanding. If that cent is more than 90 days overdue, the naïve approach will label the account as "defaulted." At the same time, banks might have a sensible operational rule that if a defaulted customer owes them less than a dollar, they don't do anything about it because the cost of chasing a few cents is a lot more than the money owed; instead, they will periodically write off these accounts. Now let's further assume that these rounding issues only occur if the loan amount contains a fraction of $12,000—this is because the bank sets interest rates with three digits after the comma, and 0.001% on $12,000 is exactly $0.01 per month (= $12,000 * 0.001\% / 12$). Can you see what will happen? A clever algorithm might figure out that the risk of default is much lower than normal if the loan amount is a multiple of $12,000—and thereby create a loophole where some risky customers might slip in and get a loan although for any other loan amount the algorithm would reject them. The data scientist therefore must review the distribution of defaulted balance amounts, identify this issue with immaterial values, and exclude any record with a balance below the operational threshold where the bank would commence collections efforts. It's important to note that unlike data cleaning (which deals with factually wrong data), this step deals with conceptual issues caused by factually perfectly correct data. This step therefore hinges a lot more on domain knowledge and judgment by the data scientist.

- *Feature development* is the process where new variables are created to tease out insights from raw data that can be used as inputs in an algorithm. In the hair example, you encountered the coding of a dummy indicating males (i.e., male = 1, female = 0) as a very simple example. The distance of a mobile phone's location right now from the closest of the top three locations where in the past 12 months it has spent most of its time is an example of a very complex feature. Maybe you are trying to combat online fraud—in this case, you might have the idea that if the mobile phone is far away from its typical location, it is more likely that the phone was stolen and a thief is trying to use it. What is a typical location? Here you have decided to define the top three locations. For that, you first need to create a log of the phone's location for, say, the past 12 months, so you need some form of location data (e.g., from log-in events when the phone connects with a cell phone tower, or possibly more precise locations recorded by an app or sent with search queries), process it to assign a location to each unit of time, and make assumptions for how you deal with gaps in the data (e.g., how you deal with a situation where for two weeks no data was sent at all—it is possible that the person did stay at home (maybe tending to a sick family member) but it is more likely that the phone was broken and sent to a repair shop, or that the person went hiking in the Himalayas—hence you may decide that 12 or 24 hours after the last signal you set the location to "unknown" and exclude time spans with unknown location from the analysis). You then need to aggregate total time spent per location, choose the top three, and calculate the distance of the phone's current location to each of the top three locations to derive the variable you had in mind. And this isn't even the most complicated variable I have ever seen! But it shows three things: it is fun, it is complicated (and hence time-consuming), and it involves a lot of assumptions and judgmental decisions. Remember that last bit—this is obviously where biases come in!

- *Short-listing of variables* is to the columns of the huge data table that data scientists work with what exclusion of records is to the rows: we delete individual variables (i.e., columns—in order to build an algorithm, we usually need to arrange all data in a huge table where each observation (e.g., a person in the sample for which you have observed the amount of hair as well as the predictive attributes) is a row and each predictive variable (either a raw attribute you have collected such as age or a feature you have calculated based on other data) is a column). It is also a step that often is skipped at the data scientist's peril. Many features considered are absolutely useless (i.e., have no predictive value at all) while others might be predictive but redundant because they are very similar to (in the statistician's language: highly correlated with) other features (as an extreme example, a person's weight in kilograms and the weight in pounds are exactly the same information, just expressed in different units). If these variables are kept in the sample, they don't have any benefit but they can wreak all kind of havoc—in extreme cases (such as two perfectly correlated variables like in the weight example) they actually can crash the model estimation procedure (which in a way is the lucky case because then at least the data scientist will notice) but usually they just play all kinds of naughty tricks on the data scientist, and some of that will cause a biased model.

- *Model estimation* is the step of actually estimating the model's coefficient—this is where we might do some matrix algebra for OLS regressions or run scripts in a statistics package to build a gradient-boosted decision tree with XGBoost (i.e., build a very complex model using a tool somebody else has developed so that we *even* could do this without knowing exactly what we're doing…). This can be a lot of fun, especially if we try out some fancy new algorithm we have never used before and the resulting model's performance is 0.0001% better than the standard algorithm we normally use!

- *Model tuning* is a series of iterative steps where the data scientist looks at the initial results, assesses them (e.g., detects an unwanted bias or other unpleasant behavior of the model), and tries to fix the problem in one of three ways: the data scientist can either select a different set of rows of the data (e.g., belatedly remove accounts with one-cent-balances), or change the columns with the predictive features (e.g., fix a logical flaw in one of the variables), or change some of the parameter settings of the model estimation procedure. The latter are also called *hyperparameters*—just how a baker can adjust the temperature and the humidity in the oven and adjust the baking time, you can think of a model estimation procedure as a machine with a couple of dials and buttons you can play around with to get better baking results. I'll illustrate this with the estimation of decision trees: decision trees can have a tendency to overfit the data (e.g., if your sample for the hair model contains three bald people who all happen to be born on March 1, an overzealous decision tree might conclude that being born on March 1 is an excellent predictor of being bald). One option to counteract this is the so-called Bonferroni correction (it basically "dials up" the number of people with the same attribute it wants to see before it believes that this is *not* random). The Bonferroni correction can be very conservative, however, and you might decide to instead try the Holm–Bonferroni method or the Šidák correction.[1] For many types of models, the possible variations to such hyperparameters are almost limitless, and sadly there is no one way to set these hyperparameters that can be considered universally correct or better than all other settings. That means that the data scientist's judgment in choosing these hyperparameters is very important—and yet another potential source for bias!

- *Calibration of model outputs and decision rules* is the step where the raw model output (e.g., a probability of default) is converted into a decision rule for business applications (e.g., whether to approve or reject a loan application).

---

[1] This book is not meant to be a text book about multiple comparisons in statistics; I merely want to illustrate that even apparently simple statistical methods entail myriad little decisions that can affect outcomes. It's the same with electricians—it sounds straightforward to ask for a power outlet to be installed in the garage but the electrician faces myriad little decisions such as whether to use the same fuse as your freezer or a different one, what rating the fuse should have (i.e., how much amperage it can carry), etc. If the fuse blows and your ice cream in the freezer melts every time your father-in-law operates a power tool in the garage, your electrician clearly has made a bad choice with the hyperparameters!

Here a lot of additional considerations and judgments come into play (e.g., approval decisions often consider some sort of profitability criterion, and that requires the allocation of cost) which entails some very philosophical and, in the end, arbitrary judgment calls. And by now you know that where there is judgment, there is bias…

- *Model documentation* is the step where the data scientist writes down what he or she has done so that others can form an understanding and independent opinion of the model. The conceptual validation and proper use of the model by others are critical to prevent biases—and hinge on an appropriate model documentation. If there are gaps or misrepresentations in the model documentation, the reader's biases will kick in (e.g., if the model documentation boasts a high predictive power, this will trigger an anchoring effect and confirmation bias that can prevent a reader from asking clarifying questions to uncover some of the hidden problems of the model).

As a result, model assembly derives an algorithm that now can be implemented in decision processes such as approving a loan, alerting airport security of a potential terrorist, or suggesting you to buy my latest book.

Step 4: Model Validation

Model validation may be informal or, as in the case of regulated financial institutions, a formal governance process executed by a dedicated unit of the organization. It is inspired by technical inspections that have proven their value in many other realms of life—for example, in many countries cars require a regular technical inspection to validate that they are still safe for use on public roads. As you will hear in Chapter 7, one root cause for algorithmic biases are the data scientist's biases, and the independent challenger function created through model validation can be an effective counterbalance to such biases.

Step 5: Model Implementation

In most cases, a model is developed on a different computer system than the system where actual business transactions are processed. In order to use the model in "real life" to make business decisions (this often is called "in production"), additional work steps are required, which are called *implementation*. For example, once a data scientist has developed a new scorecard for approving credit card applications, the bank needs to upload the algorithm in its credit decisioning (IT) system and create a process to collect the data required as inputs in the scorecard. As you will see in subsequent chapters, the way this data is created or collected from other sources—and how the system deals with missing or nonsensical values in live operations—also can introduce biased decisions; therefore, model implementation needs to be considered part of the scope of fighting algorithmic bias.

Summary

In this chapter, you reviewed the entire process of developing models; on a high level, you explored the five major steps involved in building a model:

- *Model design* ensures that the model supports its strategic objective by defining the outcome to be predicted, on what kind of population it is developed, which predictive data is used, and what modeling methodology is applied.

- *Data engineering* prepares appropriate data for the model development by defining a suitable sample; collecting raw data; splitting the sample in three parts for development, testing, and validation; ensuring high data quality by identifying and cleaning issues with the data; and aggregating granular data.

- *Model assembly* produces the actual algorithm. This step involves seven substeps, namely the exclusion of inappropriate records based on logical criteria, the development of new features, the elimination of features that are useless or redundant, an initial estimate of the model coefficients, their iterative tuning, the calibration of model outputs and decision-rules around them, and the documentation of the model.

- *Model validation* is a governance process to independently ascertain the model's fitness for use.

- *Model implementation* deploys the model in actual business operations; this involves in particular feeding data inputs into the model and linking model outputs to business decisions.

The discussion of these five work steps has introduced you to the work of the data scientist and the process required to develop an algorithm with statistical techniques. I have not yet differentiated different modeling techniques, apart from alluding to different levels of complexity. A class of modeling techniques you certainly will have heard a lot about recently is called machine learning. In the next chapter, I will unravel facts and myths about machine learning.

Machine Learning in a Nutshell

You have probably already heard about machine learning—it has become a buzzword associated with everything from utopian paradises where machine learning seems to be able to solve almost every problem in a day to scenarios where machine learning is associated with dire biases suppressing humans of all shades.

In reality, machine learning is a lot more basic—it's one of many tools a data scientist has at her disposal (it actually has been around for decades; it has just recently become both a lot cheaper to buy powerful computers capable of running advanced machine learning tools and a lot more accessible because off-the-shelf tools make simplified versions of machine learning available to anyone capable of hitting a button on a computer).

In this chapter, I will discuss what problem machine learning tries to solve, how it does so, and how this compares to other statistical techniques available to data scientists. This chapter therefore is really aimed at beginners, so data scientists familiar with machine learning can jump ahead right to the next chapter.

© Tobias Baer 2019
T. Baer, *Understand, Manage, and Prevent Algorithmic Bias,*
https://doi.org/10.1007/978-1-4842-4885-0_5

Objectives of Machine Learning

In Chapter 3, I discussed a very simple algorithm to estimate the number of hairs of a person. The algorithm consisted of a linear combination of just three independent variables. Without doubt, empirical validation of the algorithm would show that it is a rather poor predictor of the number of hairs. In order to improve the algorithm, we would look in particular at three techniques: non-linear transformations of the independent variables and other advanced derived features, differential treatment of subsegments, and adding more explanatory variables (i.e., collecting data on additional attributes of people):

- Non-linear transformations often are required to adequately describe the relationship between a predictor and the outcome. I discussed before that age doesn't have a uniform impact on the number of hairs of a person. In early childhood, we might expect that the amount of hair we have increases from the time we are born; as adults, we might expect humans to maintain a more or less constant number of hairs; and only in old age would we expect the number of hairs to rapidly decline. A quadratic transformation of age and hyperbolic cosine of age are examples of new features derived from age that depict something of a U-shape; with a U-shaped variable we can construct a relationship where "peak hair" is reached at a particular age, such as 25. For example,

$$x'_2 = (age-25)^2$$

  is a number that reaches 0 at age 25 and a value of 625 (= 25^2) both at birth and age 50. x'_2 therefore is something of an abstract feature—a kind of "penalty factor" that happens to be 625 at certain ages. What penalty do we want to assign at age 50? If we stick to our implied earlier estimate (50,000 = 1,000 * 50), we now would need to set $\beta_2 = -80$. This looks reasonable—however, what would our model suggest for a person who is 100 years old? $-80 * (100-25)^2 = -450,000$. You can easily see how for very old people the equation might suggest a negative number of hairs, which doesn't make any sense at all! In order to ensure sensible estimates for all ages, we therefore will have to do more work—we also may have to apply some flooring or capping. Sigh!

- Subsegments are groups of people within our population that require a different treatment. For example, our sample includes both men and women. It seems that men are more likely to become bald when old than women; the effect of age on the number of hairs therefore may depend on gender. A simple way of capturing this in our algorithm is a so-called interaction effect: we introduce a fourth variable that is created by multiplying a gender-dummy (0 for women, 1 for men) with age:

$$x_4 = \text{gender} * x_2$$

This has an interesting effect: the coefficient β_2 estimated for x_2 now will be the effect of age on the number of hair for women (as for women, gender is 0, also x_4 will be 0 for all women and hence simply drop out of the equation) while for men the effect is expressed by $(\beta_2 + \beta_4)$—essentially β_4 is a correction applied to β_2 for men. This might be an excellent improvement of our algorithm—but it is still crude because the gender effect might only apply to old people but not to children. Again, we may need to explore more complex corrections to solve this. Sigh!

- Additional data may be critical to improve the accuracy of our predictions. The properties of one's hair seem to be heavily driven by one's genes, so one idea could be to add the number of hairs of both the mother and the father to the algorithm. But why stop here—why not add the detailed structure of each person's genome to the algorithm? If we define one variable for each DNA base pair of the human genome, we would add about 3 billion variables. This would be what data scientists call *big data*. (Voice, pictures, and detailed data on movements also entail big data.) To be honest, 3 billion variables would be too much data simply because in statistics, we need more rows (i.e., people in our sample) than columns (i.e., variables), so we would need to work with a genomics expert to zero in on the specific parts of the genome likely to influence hair. Sigh!

These examples make it clear that improving an algorithm quickly can become an extraordinary amount of work—more than even the hardest working data scientist can accomplish within a reasonable amount of time, especially if the user of the algorithm cannot wait a long time. Machine learning tries to solve this problem.

Human history repeatedly has replaced manual labor (e.g., ploughing a field or washing clothes with a washboard) with machines that automated part of the process. Machine learning does the same for statistical modelling.

This automation and the massive reduction in time required to develop an algorithm also allows for a paradigm shift. Traditionally, building a model was a bit like shooting a satellite into the orbit—data scientists might spend many months on building a new algorithm, but this algorithm often would be in use for many years. The speed and low cost of machine learning allows for rapid updating of algorithms—if a new algorithm is estimated every month, week, or day, the improvements from one version to the next may be incremental but at every point of time, the model takes the most recent changes in the environment into account.

In fact, there are even self-improving machine learning models that update themselves without any human intervention. Real-time machine learning takes this to the extreme—for every transaction, a new algorithm is estimated that takes the most recent data into account.

A Glimpse Under the Hood of Machine Learning

The following analogy can give you an idea what machine learning does compared to a traditional data scientist's more manual approach.

Traditionally, if you wanted to know how best to drive from point A to point B, you might have asked a taxi driver for directions. The quality of the response depended critically on the taxi driver actually knowing the way—while the taxi driver still might consider several routes and make an assessment of which route at this specific time of the day would be the fastest, the taxi driver's *a priori* knowledge of one or more possible routes was a critical prerequisite for getting a sensible answer.

Machine learning is a generic technique to crack the same problem without any prior knowledge. In its simplest version, a computer could map out every possible way to drive from point A to point B (including the truly moronic routes), calculate the driving time required, and tell you the fastest route.

And whiz kids can develop more elaborate versions of machine learning by programming clever algorithms that avoid wasting time on calculating truly silly routes (thus increasing the speed with which the algorithm spits out the answer) and that possibly even optimize the route by other criteria. Sometimes the algorithm will come up with the same answer as a seasoned taxi driver, while at other times, it might come up with a stunning shortcut the driver never had considered (e.g., it may turn out that in spite of its draconic speed limit, a cut through a school zone is faster than getting stuck at an eternally congested intersection the main road crosses). And if you give the computer access to real-time traffic data, it can find the fastest way based on current traffic conditions. This is an aspect where the computer can easily outperform the taxi driver as a person sitting in a car around point A normally cannot see the current traffic along the entire way and at best can only guess traffic conditions based on past experience.

Applying machine learning to the development of a predictive model is a bit like tasking a computer with finding a driving route: the machine learning algorithm basically tries to find every reasonable way (and tons of silly ways) to make sense of the data, and gives you the best equation it could find. It therefore specifically addresses the three "fun parts" of the model assembly process discussed in the previous chapter:

- *Feature development* is aided by machine learning algorithms that automatically generate thousands of derived variables. We discussed before in Chapter 3 how the hair algorithm might achieve better accuracy if we found not only a good non-linear transformation of age but also an appropriate floor and cap. Machine learning can loop through an endless number of options, be it a seventh-degree polynomial equation, a trigonometric function, or a Fourier transformation of time series data.

- *Model estimation* is aided by machine learning estimating complex models that allow for a lot of flexibility—for example, a decision tree in effect can differentiate various subsegments and treat them completely differently. In fact, the universality theorem states that deep learning, the currently most advanced machine learning technique that uses neural networks and therefore emulates in a way how the human brain works, can approximate *every* possible function. (Just remember that "can" does not mean "necessarily will every single time" and usually comes with all sorts of conditions that very often might not be met by reality.) And machine learning may not just estimate a single model—many machine learning techniques aim to produce better and more robust results by estimating a large number of models that are used simultaneously, so-called ensembles of models (imagine a committee of highly paid expert robots who all cast a vote or estimate).

- *Model tuning* is aided by recursive machine learning techniques that specifically learn from their own errors. Previously I mentioned gradient boosted decision trees—they are an example of models tuned by machine learning. Specifically, the machine learning algorithm examines the *errors* of an initial predictive model and fits a corrective model on the errors. This process can be iterated until either we are happy with the achieved level of accuracy (data scientists tell the statistical tuning algorithm a numerical criterion for when we are "happy"—if the incremental improvement achieved by the last iteration falls below a certain threshold) or we reach computational limitations (i.e., doing more iterations would crash the computer or simply take more time than we have).

Thanks to these powers, machine learning can even use data sources that could not easily be used by more basic techniques—for example, voice recordings, images, or yes, data on the human genome.

However, that's it—machine learning automates three important tasks within the overall model development process but leaves many work steps untouched. In particular, it doesn't address many of the work steps that require business insights or hard problem-solving (e.g., deciding which records to exclude from the sample).

A Comparison of Machine Learning with Other Statistical Modeling Techniques

Unfortunately, there is a lot of hype surrounding machine learning. Some people ascribe to machine learning magical powers that solve every single shortcoming of more traditional statistical algorithms, while others believe that with machine learning, a new predictive model can be built in a day—a naïve perspective that blissfully ignores all the work steps not addressed by machine learning.

Again, the analogy with our taxi driver can illustrate some of the limitations of machine learning: in spite of all its data, a navigation computer cannot make some of the calculations a taxi driver would. For example, a taxi driver might glance at the dark clouds in the sky and figure that by the time he reaches the ferry to cross the harbor, ferry service may be suspended due to inclement weather (Google Maps definitely has failed me on this one—the taxi driver might rather head for the slightly longer tunnel route). Or he may choose to avoid a shady neighborhood that recently saw several robberies of cars stopping at red lights. Humans can take a holistic, creative approach and temper themselves with prudence whereas machine learning takes a brute-force approach to cranking out predictive models that solve exclusively a narrowly defined objective.

The boundary between machine learning and other statistical techniques actually is fluid. The term *machine learning* was coined in 1959,[1] a few years after the field of artificial intelligence was founded.[2] As soon as computers started to lend humans a hand in computations, statisticians started to use them as a way to handle computationally more demanding techniques—thus the use of maximum likelihood estimators spread and increasingly even students could develop decision trees using CHAID (a technique that is simple compared to today's tree ensembles but almost infinitely more demanding than an OLS regression). The latest boost happened recently when statisticians figured out that computer chips originally developed for creating high-quality graphics can also parallel-process statistical operations. Yes, we are talking about the gaming applications where computers depict fast objects flying through three-dimensional space—we need to thank our kids, who spent half their childhood on computer games and hence clamored for ever more powerful machines! One NVidia laptop packs as much computing power (so-called "cores" that are packed in a GPU, or graphics processing unit) as several thousand traditional laptops (which use a sequential CPU, or central processing unit). This level of computational power is in particular needed for deep learning, the right now most advanced type of machine learning.

For our purposes, I actually don't believe that a distinction between machine learning and other, more traditional techniques matters. Instead, two aspects stand out:

- Machine learning can automate several steps in the development of a statistical algorithm that historically have consumed a lot of time for data scientists. In doing so, it can help eliminate some biases—for example, where traditionally a data scientist may have carefully picked a limited number of predictors because of the substantial manual effort involved in collecting and processing each data field and through this selection of predictors might have introduced an availability or confirmation bias, machine learning promises to wade through tens of thousands of potential predictors, thus testing also the most unlikely ones and challenging the data scientist's prejudices.

- At the same time, the ease with which software packages allow even laymen to use machine learning also poses new risks. Aided by the surrounding hype, laymen as well as

---

[1] Arthur Samuel, "Some Studies in Machine Learning Using the Game of Checkers," *IBM Journal of Research and Development*, 3 (3), 210–229, 1959.
[2] J. Moor, "The Dartmouth College Artificial Intelligence Conference: The Next Fifty Years," *AI Magazine*, Vol 27, No., 4, 87-9, 2006.

rushed or naïve data scientists may use machine learning without paying enough attention to the work steps not addressed by it (e.g., data cleaning). As a result, there is an increased risk of biases slipping through—because data anomalies remain undetected.

Ironically, machine learning resembles humans more than simpler statistical algorithms in an important way: while simpler statistical algorithms interact with data scientists' logical thinking through a much more manual and therefore transparent process, machine learning emulates the human subconscious, the fast and seemingly effortless pattern-recognition machine that introduces the biases in our thinking. Just as we cannot directly observe the subconscious mechanisms causing most of our decision biases but have to detect and analyze them through psychological experiments, machine learning is similarly intransparent, and therefore again we struggle to recognize biases simply by looking at a predictive algorithm created by machine learning. Instead, we need to indirectly establish the presence of biases by analyzing outputs and behaviors of models.

Just as the industrial revolution created unprecedented opportunities for human activity and had far-reaching consequences for our lives, machine learning does promise to unlock many new applications for algorithms that could benefit humanity in many ways. Thanks to automation, with machine learning the cost of developing or updating a statistical algorithm can drop significantly. As a result, businesses have started to apply algorithms to many decision problems to which an algorithm has never been applied before, and they have started to replace existing algorithms at a much greater pace—but at times they also may let go of manual oversight and validation of the development of algorithms. As a result, thanks to machine learning, algorithms have become a lot more ubiquitous in organizations, as have the risks for algorithmic biases.

Summary

In this chapter, you reviewed a particular subset of techniques for developing statistical algorithms, namely machine learning. A couple of observations stood out:

- Machine learning enables the development of more advanced models primarily by supporting more complex features and model designs, more differentiated treatment of subsegments, and consideration of significantly more data including big data and inherently complex attributes such as pictures and recorded speech.

- Through these features, machine learning can challenge and thus debias some of the data scientist's beliefs.

- Machine learning also automates several substeps of the model development process and thereby allows the cost-effective deployment of algorithms to more and more decisions.

- By automating previously manual model development steps and tempting data scientists and users to omit other steps (sometimes out of a naïve belief that machine learning eliminates the need for all human oversight), machine learning can introduce new biases to algorithms.

- And because of its intrinsically opaque nature, biases in algorithms developed by machine learning mostly need to be diagnosed and addressed indirectly, similar to the way how human biases are diagnosed and addressed.

Armed with this solid understanding of statistical algorithms in general and machine learning in particular, we now can start a deeper exploration of algorithmic bias in Part II of this book. As a first step, the next chapter will explore how biases in the real world can be mirrored by algorithms. Subsequent chapters will also examine biases created by algorithms.

Where Does Algorithmic Bias Come From?

How Real-World Biases Are Mirrored by Algorithms

Now that you have heard about the many ways human behavior can be biased on the one hand and how complex the development of an algorithm is on the other hand, you have probably started to appreciate in how many ways algorithmic biases can arise. In this second part of the book, we will examine in greater detail the different ways algorithmic biases can be introduced.

In this chapter, we will grab the bull by its horns and tackle first the most difficult type of algorithmic bias: the algorithmic bias caused by biased behaviors in the real world. It is the most difficult type of bias because in a sense this algorithmic bias is "correct"—the algorithm does what statistically it is supposed to do: it correctly mirrors and represents the real world. We therefore are not grappling just with technical issues but with deep philosophical and ethical problems. Importantly, we will conclude that algorithms can be as much part of the solution as they can be part of the problem.

© Tobias Baer 2019
T. Baer, *Understand, Manage, and Prevent Algorithmic Bias,*
https://doi.org/10.1007/978-1-4842-4885-0_6

I earlier stated that statistical algorithms are a way to remove bias from human judgment. However, algorithms sometimes fail to deliver on this promise. The reason is that real-world biases sometimes create facts, and these facts now are the reality that shapes the algorithm's logic, thus perpetuating a bias.

To illustrate, let's examine this fictitious example of confirmation bias: if the police are more likely to frisk passers-by with green skin (a.k.a. Martians) than passers-by with grey skin (a.k.a. Zeta Reticulans) even though both populations have exactly the same share of people carrying illicit drugs (e.g., 5%), then the police are likely to finish the day with a disproportionately large number of Martians caught with illicit drugs. Assume you want to overcome this biased behavior by programming a device that scores each person passing a police officer and beeps when the algorithm suggests frisking the person because the algorithm detects a high probability of carrying drugs. If both groups have the same propensity to carry illicit drugs, you would expect your device to also seek out members of both groups with the same frequency.

In order to construct the algorithm, you might collect data on all persons frisked by the police in the last 12 months. For each person, you collect a lot of different attributes as well as the results of the frisking—a binary flag if illicit drugs have been found. Based on the information above, you might expect to find that both 5% of Martians and 5% of Zeta Reticulans have carried illicit drugs—or maybe you expect a 50% success rate because you believe that the police are really good at spotting criminals. However, you find that 20% of Martian and 10% of Zeta Reticulans in the sample had carried drugs. What happened?

First of all, you can assume that the police somehow figured out a way to target people that are more likely to carry illicit drugs than the average person because their success ratio is significantly higher than 5%—the police therefore have some real insight that you would want to capture in your algorithm. However, why is the propensity of carrying illicit drug doubled for Martians?

It is possible that for the police it is somehow easier to spot illicit drugs on Martians than on Zeta Reticulans—maybe Martians prefer tighter-fitting clothes that make it easier to spot packages in cargo pockets. However, it is also possible that confirmation bias subtly changes the police's behavior. If the police expect every Martian frisked to carry illicit drugs, they will not only be very diligent but if the first frisking doesn't reveal any results, they might frisk once more, looking for hidden pockets. On the other hand, if many Zeta Reticulans are frisked only pro forma in order so as not to give an impression of being overly biased, a police officer may just frisk a couple of pockets and then let the Zeta Reticulan go. In other words, *confirmation bias* has compromised your data because some Zeta Reticulans carrying illicit drugs go undetected because of a lighter frisking practice.

This behavior occurs in many contexts. It also happens in hiring. Due to the *anchoring effect*, interviewers often have formed an opinion of an interviewee within the first couple of seconds of an interview, maybe as soon as the candidate walks into the door. This opinion will now inform the *confirmation bias*. If the candidate gives a mumbled answer to a quiz question, the interviewer's brain might (subconsciously) "hear" the correct answer if it has already decided that this is a stellar candidate—and it may interpret the mumbling as proof that the interviewee isn't a stellar candidate if the brain already has rejected the candidate.

Even more crazily, however, the interviewee might detect the interviewer's biased judgment through body language and speech[1]—and subconsciously adjust to this. Interviewees who feel that the interviewer has a low opinion of them actually *do* perform worse. If you recorded the interview by video and developed the most advanced deep learning model on the planet to objectively score the interviewee's performance, the score would still indicate a performance concordant with the interviewer's bias even though it is entirely a psychological artifact.

This illustrates a major dilemma: human biases to an extent shape the world—and where biases have translated into factual differences in the behavior or appearance of intrinsically equal subjects, algorithms lose their power to correct the picture simply by applying statistical techniques to data.

Let's push our thought experiment a bit further: realizing that the police uses different frisking protocols for Martians and Zeta Reticulans, you could work with a group of officers to run an experiment where the police agree to follow exactly the same approach for every person they frisk—maybe even use a portable body scanner (like the one used in airports) to complement the manual frisking. To your great surprise, even though now the rates of carrying illicit drugs have come a bit closer to each other, Martians still are found to carry drugs more frequently. You start to doubt your hypothesis. What if Martians really have more criminal energy?

In reality, you might be dealing with a much deeper issue. Years of bad press (the *Zeta Evening Standard* frequently runs headlines like "Another 15 Martian drug dealers arrested," quietly ignoring the 7 Zeta Reticulans indicted the same day) may have influenced public opinion, and Martians therefore may struggle more than Zeta Reticulans to find jobs. As a result, more of them might end up dealing in drugs.

---

[1]Also, job interviews illustrate this point: If an interviewer *believes* that an applicant is ill suited for a job (e.g., because of a bias against males with red finger nails), applicants actually *do* perform worse, apparently because they pick up on the interviewer's beliefs through body language and subconsciously adjust their behavior, which once again proves that humans can be as bad or worse than biased algorithms! Read more about the impact of subconscious biases on interview and workplace performance at www.forbes.com/sites/taraswart/2018/05/21/prejudice-at-work.

Biases therefore often have a "winner takes all" effect—an initial bias starts to tweak reality, and the effect becomes self-reinforcing and even self-fulfilling. Replacing human judgment with an algorithm at this point often enough can cement the status quo. If you developed a world class algorithm with the data you have collected, you are prone to achieve a much higher success rate than the judgmental approach of the police—thanks to your algorithm, the police may end up frisking overall less people but find 80% of them to carry illicit drugs, hence significantly increasing the number of drug dealers apprehended. Yet the vast majority of people flagged by your algorithm may be Martians, and following several lurid articles about the police's work, one day the *Zeta Evening Standard* will publish the first Letter to the Editor openly musing whether the city should ban Martians.

Repeating the wrong answer thrice doesn't make it right. If you face a situation like the one we are considering here, however, it is important to correctly identify the foe you are fighting: The algorithm is *not* biased—it is an unbiased representation of a reality that is horribly flawed due to human bias. To right the situation, it therefore is insufficient to just fix the algorithm—it is necessary to fix the world.

However, what should you do about your algorithm? There is a short answer and a long answer.

The short answer is that your algorithm right now is perpetuating a biased view of Martians that fuels ever-increasing discrimination and injustice. Your algorithm has become an accomplice. As a first step, you therefore will have to consider whether you want to stop using your algorithm. This is a difficult ethical decision and not within the scope of this book. However, I invite you to imagine what in our little story would happen if you took the algorithm away from the police. Would the author of said Letter to the Editor and others thinking similarly change their views of Martians, and would they accept if the police stopped frisking Martians? Or would they demand that the police double up their fight against drug dealers, driven by an honest fear for their safety and the future of their children? And what would the police do without your algorithm? Would they start to frisk Martians and Zeta Reticulans with the same frequency and utmost neutrality, or would they revert to even worse biases than before?

Google faced a situation like this in 2016 when a reporter found that research results propagated hate speech—for example, if a user started typing "jews are" in the search engine.[2] Google found a simple solution: auto-complete now is blocked for potentially controversial subjects.[3] Unfortunately, there is

---

[2]www.theguardian.com/technology/2016/dec/04/google-democracy-truth-internet-search-facebook
[3]www.theguardian.com/technology/2016/dec/05/google-alters-search-autocomplete-remove-are-jews-evil-suggestion

not always such an easy way out if an algorithm reflects a deep bias that has crept into society.

The long answer is that your algorithm could become part of the solution. In our little example, your algorithm now wields supreme power in deciding who gets frisked. If your algorithm was changed in a way that is more "just," it would result in more just outcomes. The big issue is that here the definition of justice is outside of the realm of statistics.

In the third part of this book, we therefore will visit a wide range of options one could consider for this "long answer." There might be ingenious ways to collect better data that trumps the bias, or there might be a democratic, political process to define what the electorate considers "justice" as a basis to inform a management overlay over the algorithm in order to drive actual decisions.

Summary

In this chapter, you explored situations where an algorithm dispassionately mirrors a deep bias in society. Key insights are:

- If biases in the real world have created their own reality, statistical techniques lose their power to remove such biases on their own.

- In such situations, an algorithm arguably becomes an accomplice that can perpetuate biases and thus entrench ramifications deeper and deeper into reality.

- Nevertheless, even a biased algorithm might be the smaller evil compared to human judgment that implies even worse biases.

- In the long term, algorithms can even be the solution to real-world biases, which we will discuss in detail in part III of this book (in particular in Chapter 16).

In a way, you have made it through the darkest section of this book. There are many other ways algorithms can be biased, but they all tend to be easier to solve than the particular issue discussed in this chapter, and the many solutions at our disposal will be the subject of the third and fourth parts of the book. The sun therefore now will slowly rise again, and with each chapter you hopefully will feel more empowered to make the world a better place by understanding, managing, and preventing algorithmic bias.

Data Scientists' Biases

In the last chapter, we considered the most dire situation possible—biases that have been so deeply entrenched in reality that it's impossible to collect data to refute them. Very often, however, there is the data required to keep biases out of the algorithm—but somehow the data scientist lets a bias slip through nevertheless. This chapter looks more closely at this cause of algorithmic bias.

When we discussed the model development process in Chapter 4, you learned that the data scientist not only needs to go through a lot of work steps but that many steps also require judgment. There are three major sources of algorithmic bias introduced by the judgment of data scientists:

- *Confirmation bias* that sets up the model to replicate a bias in the data scientist's own mind;

- *Ego depletion* that distracts the data scientist from opportunities to avoid bias; and

- *Overconfidence* that causes the data scientist to reject signals that the model might be biased.

Confirmation Bias

It's a common wisdom that you need to ask if you want to know something—and data can be like this too. If you have a strong hypothesis and don't ask

© Tobias Baer 2019
T. Baer, *Understand, Manage, and Prevent Algorithmic Bias*,
https://doi.org/10.1007/978-1-4842-4885-0_7

your data about alternatives, it may not tell you. Model design and sample definition (the first two steps in the development process) can set up a data scientist for a fundamental confirmation bias by excluding data from the modelling effort that would tell a different story.

Let's first explore how confirmation biases can compromise model design and then turn our attention to sample definition.

Confirmation Bias in Model Design

Model design defines what the model predicts with what data. In other words, you decide what question you ask the algorithm. If you ask a biased question, you are bound to get a biased answer!

One important aspect of model design is the dependent variable—how to actually define good and bad. I once served a Chinese bank that wanted to solve a standard problem—they wanted to turn around a loss-making credit card portfolio by introducing a better credit scoring model. Their hypothesis was that risky customers lose money while safe customers generate profits; they therefore asked me to develop a risk score to separate risky from safe customers.

The model performed very well on the task given—it was so good in separating good from bad customers that it could reduce the number of defaults by more than half while losing less than 10% of good customers. When we analyzed the profit achieved by the new model, however, we were in for a surprise: the portfolio still lost money. It turned out that the original question asked ("Tell us which applicants have a high risk.") was a bias against risky customers. In reality, some of the risky customers were highly profitable because they not only had a considerable chance of defaulting but also paid a lot of interest and fees to the bank—enough to not only cover the expected losses but also make a tidy profit.

On the other hand, the bank lost a lot of money on many safe customers. This was because many "safe" customers were safe simply because they never used their card—pushy, commission-hungry sales people had talked many people into applying for a credit card that they didn't actually want or use. These cards generated zero revenue for the bank but still caused large sales commissions and considerable operational costs.

We therefore decided to ask the algorithm a different question: "How much money will I make from this applicant?" This was a complicated question, and we broke it down into two subquestions: one algorithm estimated the credit loss, while a second algorithm estimated the revenue from the customer. The loss-making customers comprised two subgroups: customers with scant chances of making any revenue (many of whom looked very "safe") and extremely risky customers whose credit losses would exceed any revenue we could expect to make with them. By contrast, profitable customers tended to have medium risk, as these customers were enthusiastic users of credit cards and hence generated substantial revenue, enough to cover the medium-sized losses of those who defaulted on their debt.

It turns out that many businesses have only a partial understanding of where exactly their profits come from. The question "what is good?" sounds philosophical but actually is central for an algorithm that aims to classify applicants between "good" or "bad" or assign a probability of good to each of them. If data scientists follow their biases in defining good and bad, they are bound to replicate this bias through their algorithm. The same point can be illustrated with search optimization. Here algorithms may be trained to optimize number of clicks, even though having a smaller number of higher quality clicks (e.g., customers who are going to spend a lot of money on a brand or product category) might afford much better business outcomes.

Users and business leaders in general are prone to falling into this trap as well. There is a strong bias for humans to judge complex issues (e.g., whether a person would be a good employee, spouse, or consultant) with a simple proxy (e.g., if the person has a Harvard degree). In fact, the more complex an issue, the greater the likelihood that all stakeholders involved eschew a thorough discussion and intellectual penetration of the issue. Known as Parkinson's Law of Triviality or the bike-shed effect, it was neatly illustrated by Cyril Northcote Parkinson's discussion of a finance committee that approved a mega-investment in a nuclear power plant in 2½ minutes and then had a long and animated discussion about a new bike shed.[1] This bias actually illustrates how strong nature's penchant for efficiency drives the way we make decisions. If you think of a very complex, big decision *you* will have to make soon (or so far have artfully managed to avoid), you probably immediately feel a surge of reluctance to touch it because of the tediousness of the thought process—which simply is nature telling you that you'll burn a ton of energy on this and maybe should try to get by without going there. Essentially it is the same reflex as taking a short-cut through a hedge if following the official pedestrian path would entail a huge detour.

The other dimension of confirmation bias in model design is the selection of the explanatory data. Most sins here are sins of omission—if a data scientist is biased towards a particular set of predictors, he or she is more likely to exclude alternative predictors in the sample and hence never can challenge his or her biases.

For example, let's consider a CV screening algorithm that is used to automatically identify the most promising candidates to short-list for a new opening in the data science team. It is a common belief that a degree from an

---

[1]C. Northcote Parkinson, *Parkinson's Law, or the Pursuit of Progress*, John Murray Publishers, 1958.

Ivy League university is a good predictor of professional performance. But what if this bias towards Ivy League universities is wrong?

Purely for illustrative purposes, let's assume that a different factor explains the difference in average professional performance between Ivy League students and others, namely whether or not the student has learned Latin in high school. Because of the *confirmation bias*, the data scientist may not even think of collecting this information. Without Latin language skills flagged in the database, the algorithm has no chance to point out that Latin is a key driver of professional performance. If a model is misspecified (i.e., the correct predictors are missing), the statistical algorithm will try to find the best prediction with whatever is available—and for this it will usually use variables that are correlated with the missing features. In our example, we could assume that only few kids learn Latin; Latin skills may indicate that the parents made above-average efforts to give their kids the best possible education and that they possess both good language and analytical skills (i.e., both right and left brain are gifted). All these attributes of course will also increase the probability of the kid attending an Ivy League university; hence the ranking of the university becomes a proxy for Latin skills.

The result is a biased algorithm—it will favor applicants from an Ivy League school even if they don't even know what Latin is, while it may reject applicants from less renown schools even if they know Latin so well that they can translate Virgil poems on the fly.

In my experience, *confirmation biases* often dramatically narrow the range of data considered by data scientists. For example, in credit scoring, *status quo* bias has caused an excessive focus on a few "text book" data sources such as credit bureaus. *Social biases*—especially the fear of ostracism—at the same time discourage data scientists from proposing unconventional data sources even though sometimes outright funny variables would be a lot more predictive. For example, in Taiwan many algorithms still use financial ratios extensively to assess the credit risk of companies even though bankers privately will tell you that they would not even want to lend money to a customer who does *not* know how to make a balance sheet (a pretty fictional affair) look pretty for the bank (as such a customer obviously lacks basic business knowledge). On the flipside, when a Taiwanese relationship manager suggested to a team of data scientists that he can assess the quality of a borrower through a round of golf, he was laughed at—although he had figured out a great insight: observing the client cheating during the golf game was a great predictor that the customer would also be unreliable in his business dealings and hence had a high risk of default.

Confirmation Bias in Sampling

A major source for algorithmic bias is a choice of an overly short time period for the sample to span. For example, it is a *stability bias* to assume that the past year is representative for the future. Testing an algorithm over multiple years—for example, in the case of a credit score, over periods enjoying a growing economy and periods suffering a recession—allows the statistical process to better test the data scientist's hypotheses and call out biases.

I recall a situation that illustrates this pitfall well. Prior to the global financial crisis of 2007, someone had built a credit scoring algorithm for mortgages that took recent house price increases in the area as an input—a fundamentally reasonable hypothesis and sensible variable. The data scientist who built the model was under the impression, however, that house prices will *always* rise—a great illustration of an availability bias that may affect a young data scientist who has never empirically observed falling house prices. Because of that bias, the data scientist applied a mathematical transformation to the change in house prices that did not even work with negative changes in (i.e., falling) house prices. Depending on technical choices made in implementation, such a model would either throw up an error (which would raise a warning flag) or assume flat house prices (or possibly even some positive growth) if house prices fall. In other words, the algorithm could end up "assuming" that house prices don't fall even if evidence to the contrary was fed into the equation! The initial bias of the data scientist therefore had created an algorithm with a biased, selective perception of reality just as in the introduction you have observed confirmation biases to cause selective perception in human communication.

I should note that this doesn't need to be so. The laws of physics are examples of well-designed algorithms that often can be extended to dimensions not empirically observed during their development—for example, rocket scientists are able to use formulas developed based on experiments on Earth to correctly calculate trajectories of rockets into the Earth's orbit and even towards other planets such as the moon and Mars. However, as you will see in a later chapter, designing an algorithm that will work well also outside of the range of values observed for its input variables in the sample typically requires special care and effort by the data scientist.

The sample can be biased in many other ways. Limiting it to a relatively short period representative of just one particular economic regime is one example; focusing it on a particular population segment is another example. Here *availability biases* can be devastating. You probably heard that hard drugs are addictive. This belief was based on mountains of academic research. What a surprise then when researchers conducted a longitudinal study where they tracked participants over several decades—and found that most drug users at some point stopped using drugs simply because they decided to do so (e.g., because they wanted to take on a job or get married or raise a child)![2] How could that be?

It turned out that most research on drug addicts had been done with the population most easily available to psychologists—the *clinical* population of drug addicts that are *not* able to get rid of their addiction by their own volition and therefore turn to psychiatric or medical help. It turned out that because of the *availability bias*, much research on drug addiction had been conducted on a tiny subset of drug addicts, thus causing a very biased perception of the problem.

A major foe in model development therefore is confirmation bias. In the following section, you will see how a time-dependent effect called *ego depletion* can aggravate confirmation bias and other biases.

Ego Depletion

When hearing about biases of data scientists in general, most data scientists readily concur that they have observed *other* data scientists indeed suffering from these biases all the time. At the same time, they tend to believe that they themselves don't suffer from these biases. *Overconfidence* of course plays a role, but they often will be able to point out examples "proving" that they have taken measures to avoid the pitfalls described here. One might explain this away by suggesting that it's the *availability bias* coming to the rescue of the *confirmation bias* (it's easier to remember a situation where you have introduced a new data source that gave your algorithm a big lift even though your boss had derided you when you first mentioned this data source to her, than it is to remember a situation where you have not introduced any out-of-the-box data source).

However, often something a lot more subtle and vicious is going on: our brain is not a mechanical robot but a dynamic machine. When we start a complex, cognitively demanding task such as developing an algorithm, we tend to give it our full attention. As soon as 30-60 minutes into the task, ego depletion will set in—our minds start becoming fatigued. This effect originates in nature's frugality—because our conscious, logical thinking is so energy-consuming (accounting for 20% or more of our total calorie consumption), Mother Nature doesn't want us to burn all of our mental energy on just one task. Our predecessors were better off when they rotated their attention between hunting food, finding a mate, watching out against enemies, and worrying about shelter from nature's forces. Just as after a short while of displaying the same picture your mobile phone's screen might go into energy saving mode

---

[2]www.psychologytoday.com/us/articles/200405/the-surprising-truth-about-addiction-0

and darken a bit, ego depletion will gradually dim your wits—your brain will start taking shortcuts.

There are many shortcuts at the brain's disposal; in particular, it will skip opportunities to retrieve contradictory information (*confirmation bias*) and use one easy-to-retrieve metric as a proxy for harder-to-assess metrics (*anchoring effect*). Very often, the brain will gravitate towards whatever "default decision" presents itself.

Car buyers' behavior can illustrate this. In some countries, such as Germany, buyers of new cars can customize dozens of attributes. When they start the configuration process, they agonize over the car's color and the fabric of the seat and whatever minute options are given for the first couple of items. The further they get, however, the more *decision fatigue* (the specific type of ego depletion triggered by decision-making) sets in—and as a result, the more likely they are to go with the default decision suggested by the car company.

Also, where professionals make complex assessments as a basis of decisions, they often have a "default decision" such as the "safest" option when risk is involved. Detailed statistical analyses of decision quality can reveal this. For example, when I analyzed decisions of credit officers for a small business portfolio, I found that the approval rate of otherwise identical applications dropped by 4%-points between the start of a session (in the morning or right after lunch) and the end of the session (when the credit officer would take a lunch break or go home). Doctors have been found to prescribe more often unnecessary antibiotics,[3] judges have been found to be more likely to reject parole requests of prisoners,[4] and investigators have been found to be more likely to conclude that a fingerprint found at the crime scene does not match the suspect's fingerprint.[5]

Data scientists often have to make a mindboggling number of decisions as well. In model design, they often have to sift through dozens of data dictionaries with thousands of data fields and decide which fields they want to collect because for operational or economic reasons (cost) they cannot get all data fields. While they might naturally agonize about the first dozen fields, they soon will find themselves making the include/reject decision intuitively in a split-second. While they might believe they can do this because of their vast experience and intimate knowledge of hundreds of similar algorithms (*overconfidence*), their subconscious will simply activate *anchoring* and scan

[3]J.A. Linder, "Letter: Time of Day and the Decision to Prescribe Antibiotics," *JAMA Internal Medicine*, 174(12), 2029–2031, 2014.

[4]S. Danziger, J. Levav, and L. Avnaim-Pesso, "Extraneous factors in judicial decisions," *Proceedings of the National Academy of Sciences of the United States of America*, 108(17), 6889–92, 2011.

[5]T. Busey, H.J. Swofford, J. Vanderkolk, and B. Emerick, "The impact of fatigue on latent print examinations as revealed by behavioral and eye gaze testing," *Forensic Science International*, 251, 202–208, 2015.

fields for a tiny list of two or three meta-attributes. If you have ever caught yourself including or excluding a whole swath of fields only to think "Hold on a minute; this doesn't make sense, Why don't I exclude/include this one?" it's probably a case of your conscious mind having caught the subconscious mind making an excessive number of shortcuts.

The story repeats itself in data engineering. Checking each variable's descriptive statistics for signs of data quality issues, treating outliers and missing values, and deciding on transformations all quickly move from exceptional thoughtfulness to very templatized, intuitive decisions.

And in model estimation, ego depletion will entice data scientists to stick to whatever default values are provided for hyperparameters by the script they use or by their company's guidelines.

As a result, sometimes a specific bias affecting an algorithm can arise because an otherwise excellent data scientist lets the seed for the bias slip through in a moment of mental fatigue. Unfortunately, this effect is not widely recognized or accepted. At least cars and trucks have started to feature technology to detect the driver's fatigue; maybe at some point in the future laptops will encourage their brain-taxed users to take a break in the park when they observe signs of ego depletion.

Overconfidence

I showed earlier how statistical procedures in general and the concept of significance specifically can help to remove bias—but only if they are listened to. *Overconfidence* means that sadly, such warning signs are often ignored. There is a cartoon by Randall Munroe[6] about an empirical study that tests if any of 20 different colors of jelly beans have a significant impact on acne: the first 19 flavors all are insignificant at a 95% confidence level but bingo, the 20th flavor masters the hurdle. Does this really mean that eating green jelly beans causes acne? As you may recall, a 95% confidence level means that there is a 1:20 chance that an insignificant attribute will appear significant; hence if the hypothesis holds that *none* of the colors has any impact on acne, we still should expect one of the 20 to pass the test.

The major accomplice of overconfidence is the human *need for consistency*: The human mind doesn't like changing beliefs, presumably because constantly re-evaluating decisions one has made already would waste a lot of mental energy. When we are confronted with evidence challenging our beliefs, we will often more readily come up with reasons why most likely the warning indicator is broken ("Oh, the validation sample is biased/too small/too recent/ too old.") than accepting that we've been wrong all along.

---

[6]https://xkcd.com/882/

A strong manifestation of this overconfidence bias is the behavior I like to call "torturing the data until it confesses." If an initial set of features doesn't yield a good predictive model, some data scientists are more likely to tweak model estimation ("Let's try a different modeling technique.") or create additional transformations of the same raw data than fundamentally revisiting whether the dependent variable is (conceptually or computationally) flawed or the hypotheses on the outcomes' real-life root causes are wrong.

Many data scientists will obviously strive to avoid such overconfidence and diligently investigate any sign that something is amiss. Even they could, however, still be afflicted by a weaker manifestation of overconfidence: mistaking an absence of warning signs as an absence of any problems. As a result, an algorithm that at first glance looks just right (i.e., it has a good but not suspiciously excessive predictive power and no obvious shortcomings) might get only limited scrutiny, making it more likely that hidden biases in the algorithm go undetected.

Summary

In sum, you therefore need to recognize that the data scientist's work involves a dazzling number of decisions, from hundreds of repetitive microdecisions to dozens of fundamental decisions that may dramatically change the shape of the algorithm. As a result, algorithms are heavily exposed to the data scientist's biases:

- *Confirmation bias* can affect both model design and sampling.

- In model design, confirmation bias can compromise both the choice of the *dependent variable* (i.e., the definition of the outcome the algorithm is asked to predict) and the choice of the *independent variables* (i.e., the features used to predict the outcome).

- In sampling, confirmation bias can cause the choice of an incomplete sample that lacks those observations that would challenge the data scientist's hypotheses.

- In general, confirmation bias therefore most often manifests itself in the *omission* of data or features.

- *Ego depletion* is mental tiring that can be caused by having to make an excessive number of microdecisions (or simply working for hours on the same task) and gradually introduces or increases biases as a way to minimize cognitive effort.

- Because of ego depletion, harmful biases are most likely to affect the data scientist's work in a state of mental fatigue.

- *Overconfidence* causes the data scientist to reject signals that the model might be biased even in the absence of ego depletion.

Even if a data scientist avoids any such biases, however, the algorithm may still be biased if the material itself—the data used—is biased. In the following chapter, you therefore will explore biases in data.

How Data Can Introduce Biases

An old adage is "garbage in, garbage out," suggesting that if you feed garbage data into even the fanciest algorithm, you will end up with garbage as output. The same could be said about biases. In this chapter, we will review the many ways deficient data can introduce biases into an algorithm. As you will see, some of these issues can be addressed by the data scientist; other issues actually need to be addressed by the individuals who in one way or the other generate the data (e.g., an insurance underwriter processing applications or a programmer updating a webpage).

Overview of Biases Introduced by Data

There are at least six different ways data can introduce biases in an algorithm. It is important to differentiate these types of biases because they are generated at different stages of either the real-world process the algorithm is trying to describe or the model development process, and hence the solution for preventing or removing each type of bias will be different as well.

First of all, we can distinguish two types of situations where the **process of collecting information** (and thereby creating the data we use for our modeling) introduces biases:

© Tobias Baer 2019
T. Baer, *Understand, Manage, and Prevent Algorithmic Bias,*
https://doi.org/10.1007/978-1-4842-4885-0_8

- *Subjective qualitative data* created by humans such as restaurant ratings is naturally biased; a particular aspect of this problem is how the way such data is generated (e.g., the process for assigning a rating) can create specific biases.

- *Seemingly quantitative data* entails numbers that are generated by a similar process as subjective data (e.g., an application form where the field labeled "income" is filled out by a sales person) and therefore can by affected by the same issues even though it looks deceptively objective.

Second, we can distinguish two types of situations where the **source of our data** imparts a bias:

- *Data mirroring biased behavior* is superficially objective yet will still be biased. For example, annual bonus numbers appear to be an objective metric of employee performance but still can reflect deep gender biases in the way how the organization evaluates men and women.

- *Traumatizing events* are one-off events that are not predictive of future outcomes but nevertheless create an outsized bias of the algorithm.

And third, there are two ways data scientists can introduce new biases through the **model development process** itself:

- *Conceptual biases* are biases created because of specific model design decisions that create a distorted representation of reality in the sample.

- *Inappropriate data processing* creates biases through flaws in the way data is cleaned, aggregated, or transformed.

Biased data is one of the biggest sources of algorithmic bias. It therefore merits a closer look at these six issues.

Biases Introduced by Subjective Data

In Chapter 6, you encountered real-world bias in its most dire form—where it affected outcomes (i.e., the dependent variable your algorithm tries to predict). Even where outcomes are unbiased, however, we still might find bias in some of the inputs of the algorithms (i.e., the independent variables we use to make predictions).

For example, commercial credit assessment often entails judgmentally assessing qualitative aspects of a company such as management quality. While the outcome (whether or not a company in the modeling sample has defaulted) is objective and unbiased, the assessment of management quality can exhibit the full gamut of cognitive biases. Many banks ask their staff to indicate management quality on a scale such as "very good," "good," "satisfactory," and "weak." When sales people assign this rating, they naturally are driven by confirmation bias and an interest bias (they want the application to be approved) and unsurprisingly, we find most customers to be rated "very good." The only exception I encountered was an Australian bank where sales people had an additional option: "exceptionally good." This was very fortunate because it turned out that most Australian management teams are, in fact, "exceptionally good," at least in the eyes of the sales staff.

The story doesn't end here, however—the bank I'm referring to was manually reviewing such credit applications, and credit officers would routinely "correct" such enthusiastic assessments. But how? One credit officer explained: "I simply look at the financials, and then it is quite clear what the actual management quality is!" Quiz for you: What is the credit officers' bias called?[1]

If such biases render a variable useless, the algorithm of course would reject it as insignificant. In many cases, however, the bias merely reduces the power of the variable, and hence such biased assessments still can enter and therefore affect the algorithm. Partially biased values can occur if some staff members are more biased than others or if ego depletion (introduced in the previous chapter among data scientists' biases) triggers biased assessments. In fact, in my own research I found evidence for credit officers changing their assessment behavior when mental fatigue set in, suggesting that data they collect may exhibit biases in particular if they hadn't had a break for two or more hours.

Also, drop-down fields ("feature columns" in Google terminology) and other structures of qualitative data introduced by the tools used to generate or collect such data can introduce biases. A massive bias can be introduced through groupings. For example, imagine a Zeta Reticulan bank in our imaginary world that captures the "sociodemographic" segment by assigning a label, but cruelly groups Martians together with drug addicts and criminals into a category labelled "misfits." If other categories are highly predictive, this unfortunate variable is likely to enter the algorithm but now would assign a real penalty to Martians, who therefore will struggle to obtain credit.

An example of a more subtle nudge triggering biases is non-gendered labels for professions (as they are common in many languages). This could bias the classification of borderline cases when staff manually maps a free-text answer

---

[1]Indeed, it is anchoring. In essence, the credit officers deleted the sales staff's assessment of management quality as an independent variable and replaced it with a new variable they had created that basically mirrored financial statements; it therefore arguably was a redundant variable that did not improve their assessment of the company.

into a category. Imagine how "physiotherapist" might be mapped into an imperfect categorization system where only "medical doctor" and "nurses" are available as medical professions—conforming with gender stereotypes, the majority of male physiotherapists might end up being grouped with medical doctors and surgeons while for their female counterparts, the majority may end up being considered nurses.

Biases Introduced by Seemingly Quantitative Data

You have just seen how subjective judgments on qualitative labels can be biased. Would quantitative data such as an applicant's yearly income by nature be free of such cognitive biases?

Often such data fields' theoretically objective definition is deceiving: almost nobody knows their actual income. Do you remember exactly how much was left from last year's bonus after tax? Do you keep track of all other income items, including interest you receive, fringe benefits such as meal vouchers received separately from your salary, and all the tips your happy customers give?

As a result, when such information is collected, very often answers are estimates—and thus biased. And because answers are biased, they also invite interest biases: of course many applicants have an interest in getting a large loan and therefore wish their income was a bit higher than it actually is. Dan Ariely has conducted extensive research on dishonesty and found that humans lie systematically. We pace ourselves and tie our lying usually to little stories that "explain" how this unfortunate "error" in our calculations occurred in case we get caught.[2] As a result, numbers might get rounded up, and a month where income was above average suddenly might become the norm (and yearly income is swiftly estimated by multiplying that banner month's income by 12).

The bottom line is that the way data is collected can invite biases. Such data usually is still *directionally* correct and therefore still can enter an algorithm as a significant factor. If the same data collection process continues to be used after implementation of the algorithm, the biased inputs can end up biasing predictions (e.g., professions with a lot of tips may be particularly prone to inflating their income estimates; presumably tax office clerks and accountants with fixed salaries would produce more truthful estimates of their own income and as a result might get penalized by the algorithm).

---

[2]Dan Ariely, *The Honest Truth About Dishonesty: How We Lie to Everyone—Especially Ourselves*, HarperCollins, 2012.

Furthermore, once the algorithm establishes a very direct link between the data input and a desirable outcome (e.g., access to a loan), interest biases can be triggered. I once looked at a histogram of income data collected by sales staff and found four discrete spikes in values. Why was it that so many people earned exactly 500, 800, 1200, or 1700 currency units?[3] I discovered that the bank's existing algorithm had a step function—applicants received 0 points for income below 500, 20 points for 500-799, 50 points for 800-1199, and so on—and the sales staff apparently figured out the process and whenever the true income was slightly below a threshold, they generously "rounded up" to give their applicant a little lift in the point score.

Biases Introduced by Traumatized Data

This aspect of a bias caused by the source of the data also merits a deeper discussion. In psychology, we know that if a child suffers a traumatizing event, the child's behavior can be affected for the rest of her life. For example, if a child is bitten by a dog, she may develop a lifelong fear of our four legged friends. The same can happen with data.

Imagine you are still working on a credit card application scorecard. You have collected historical data comprising all credit cards originated (i.e., sold or opened) in 2016 and tracked which credit cards defaulted in the 12 months subsequent to origination. In early 2017, unfortunately there was a major natural disaster in the country's predominant region for citrus plantations; a severe hurricane and subsequent flooding brought life to a halt, destroyed most of the citrus plantations as well as thousands of houses and commercial buildings, and even forced your bank to temporarily close many of its branches. Naturally many people were unable to pay their credit card debt—some simply didn't have access to a check book for two months, while others saw the source of their livelihood destroyed and were destitute even a year after. How would this affect your algorithm?

Your data naturally would have marked many or maybe even most of the credit cards originated in this region as defaulted. In that sample, any hint that an application originated from the region—for example, an indication that the applicant works in the plantation industry or, if the jurisdiction in question allows it, even the postal code—would be indicative of a high probability of default. The disaster therefore would be deeply burned into the algorithm's logic. What would the algorithm do if next year an applicant from said region applied for a loan? Like a person with a childhood trauma, the penalty effect would live on in the algorithm's decisions, and the algorithm implicitly would assume that the disaster in this region never stopped and hence may reject most or all applications.

---

[3]Numbers are disguised and therefore only illustrative.

Other examples of one-off events that can traumatize data are big fraud schemes (which often are either committed out of a particular branch or through a particular channel) and technology-related issues that compromise a chunk of data (including cyber-attacks that affected a particular subgroup of customers only).

A related issue is so-called **outliers**. While events affect many cases at the same time (which gives the event so much heft), outliers are individual cases that are so far off the norm (e.g., a state-owned monopoly that accounts for 80% of all revenues in a particular industry) that they have an outsized impact on the equation. Statisticians also call such outliers *leverage points*. In effect, leverage points bias the algorithm towards this one case (if you like flowery labels, you can call this bias the Sunflower Management bias—it normally refers to humans who are biased towards the views of the person highest in the hierarchy, as this person will attract their attention similar to how the sun attracts the faces of sunflowers). Behind this effect is the way statistical algorithms quantify estimation errors (which they try to minimize)—if a single case has numerical values that are very, very far away from the average, this distance gives them extreme leverage in the calculation of the total estimation error, and the algorithm basically says (and I exaggerate quite a bit for dramatic effect), "We can't get this one wrong, regardless what the other data points say!"

For example, consider in your hair estimation from Chapter 3, if somehow a gorilla man (i.e., someone with a *lot* of hair) got into your sample. Let's say he speaks Italian, and a very sophisticated algorithm lumps together your gorilla man with a couple of other Italians. This creates a new prejudice that Italians have a *lot* of hair—and voilá, another mythical bias is born!

Conceptual Biases

Have you ever photocopied or photographed a page in a book in order to read it later, only to find that you accidentally cut off a part of the page and therefore missed out on an important part of the story? Conceptual biases do the same to data. Somehow the model design accidentally "cuts off" a part of the data, and the resulting gaps in the data end up biasing the algorithm.

Just as you can cut off a picture on the top or bottom, left or right, data can be cut off in different ways. The three ways to cut off data are rows, columns, and time.

Rows are cut off if some instances (observations) in the real-life population are systematically omitted. In the previous chapter, you encountered the example of research on drug addicts where, due to the researchers' availability bias, drug addicts outside of the clinical population had been ignored.

In many cases, however, the entire population is available in a database (e.g., the bank's account system) but instances are still lost in the process of extracting data from a database because of a conceptual flaw in the logic of the query. For example, if you start your sample creation with all credit cards active *today* and then proceed to filter out those that originated 1-2 years ago, you systematically miss out on credit cards that were originated 1-2 years ago but have been cancelled since.

Columns are affected if certain independent variables are missing or compromised. A typical problem is overwriting historical values with the most recent value. For example, I once worked with a start-up company that sold specific data fields from a person's social media presence such as the person's Facebook status or tweets. The company was keen on me testing their data as an input for credit scorecards; however, I realized that they had no concept of historical versus current values. They didn't archive any data and therefore could only give me a person's current Facebook status and the five most recent tweets. In order to use this data for a credit score, however, I don't need to know that they announced a week ago "just moved back in with Mom." Instead, I need to know that 23 months ago (when they applied for the loan I gave them and subsequently had to write off), they had blasted into the social media world that they "just bought the coolest house ever." Overlooking this particular issue can be lethal—updated data fields can introduce so-called *hindsight bias* (i.e., my algorithm uses information only available after the outcome). Pointing out that customers who will inform their friends in a year from now that they just moved back in again with Mom are more likely to default on any credit card I give them today between now and then not only is rather obvious but also useless. As I don't know what the person will tweet in a year's time, it's impossible to use this information for credit scoring.

Time comes into play in tracking behavior over time. Problems arise if at the beginning or end of the sample observation periods are truncated. For example, in credit modeling, some data scientists will consider all loans originated in a certain time period (e.g., from January 2010 to December 2017) and simply mark for each loan whether there has ever been a default event. How might this be a problem? For a five-year loan originating in 2010, you have the entire life of the loan in the sample and therefore implicitly have tracked the loan for five years until it was either repaid or written off in 2015. For a five-year loan originating in December 2017, however, you have only a very short history—so if you did this work in December, 2018, your so-called *performance period* was just one year. John Maynard Keynes reminds us that "in the long run, we are all dead"—so if we track loans over five years, of course we observe more default events than if we observe the same loans over just 12 months. As a result, the sample now suggests that loans originating in 2010-2013 are a lot riskier than loans from 2017. Anything correlated with the year of origination now also becomes predictive of risk because it is a proxy for the time of origination—and a new bias has been born in the data.

Biases Introduced by Inappropriate Data Processing

Finally, the maybe most tragic situation arises if an originally unbiased sample becomes biased because of some inappropriate data processing.

One example is biased data cleaning. We earlier discussed the issue of "immaterial defaults" where because of a rounding issue, a tiny amount such as $0.01 remains in the accounting system as loan balance after the customer has made the last installment, and after 90 days this "overdue" balance is flagged as a default. It is good practice to eliminate such immaterial defaults from the sample. But could there lurk a little devil in the details?

Indeed, yes. Interest calculations are deterministic. Therefore there will be certain combinations of loan amounts, origination date, and term of the loan where if the customer makes all payments on time, there will *always* be a $0.01 balance remaining at the end. If the data scientist "eliminates" immaterial defaults by literally deleting the records, she would end up deleting all good accounts for certain combinations of loan characteristics—and the only loans with the same characteristics left in the sample are true defaults. As a result, these characteristics now would become a "perfect" predictor for some of the defaults.

In this example, it would be easy to "eliminate" immaterial defaults instead by overwriting the good/bad indicator with the "repaid" label. Sometimes, however, data scientists find themselves between a rock and a hard place, and finding a data cleaning approach that does not introduce some form of bias can be a formidable challenge.

A totally different type of failed data cleaning is the so-called *silent failures* type. This occurs if an outdated table is used for mapping or transforming certain data. For example, you might be developing an algorithm to automatically score CVs of job applicants. The name of the university the applicant attended obviously could be an important variable (notwithstanding our earlier musings that it may just be a crude proxy for something else), and you may convert it into a numerical value by using a university ranking supplied by a third party. This introduces a data preparation step where the university is mapped into its rank (which is an ordinal number).

Once this algorithm is implemented, such tables should be regularly updated, especially if the rankings change over time—otherwise, your algorithm would have an eternal bias in favor or against certain schools even though over time these schools could move up or down the ranking and possibly even switch positions.

Summary

Data turns out to be a real minefield, with possible biases arising from no less than six effects:

- Some data is *qualitative* in nature and is therefore typically *subjectively created* by humans; this introduces biases into specific variables.

- Some data is quantitative or otherwise objectively defined but the process to establish its value nevertheless entails subjective elements; such data likewise can reflect human biases.

- Some data is both quantitative in nature and collected through an objective process with complete integrity but the measured values still reflect a bias in an underlying process or phenomenon.

- Some data is affected by a *traumatizing event* that introduces a bias into the data.

- Conceptual flaws in the sampling approach introduce biases by systematically omitting certain rows, columns, or time periods in the sample.

- Inappropriate data processing introduces biases through statistical or numerical artifacts.

Alas, there is more. Even if the data is perfect and the data scientist has successfully avoided all six dangers, algorithmic biases can arise. Next, we will discuss the issue of an algorithm's implicit stability bias.

The Stability Bias of Algorithms

In our review of human cognitive biases, *stability biases* featured prominently. It turns out that humans are not alone: algorithms also often exhibit stability biases. In this chapter, we will review the most important ones and explore what kind of context attributes promote them.

We first will examine the fundamental cause of the problem, namely system instability—without it, there would be no harmful stability bias. After that, we will discuss the conceptual dilemma that it is close to impossible to teach things to an algorithm that don't exist in the data, and we will explore how a slow response speed can crucially diminish the ability of algorithms to overcome stability bias through rapid learning. Finally, we will explore another dimension of the stability bias—namely in defining outcomes as good or bad.

System Instability

For any given sample, algorithms by design are unbiased (you may remember the concept of BLUE from Chapter 1). But what happens if things change? One fundamental limitation of algorithms is their implicit assumption that

© Tobias Baer 2019
T. Baer, *Understand, Manage, and Prevent Algorithmic Bias*,
https://doi.org/10.1007/978-1-4842-4885-0_9

whatever relationships existed in the sample also will prevail in the future—just as the stability bias of humans also is grounded in each human's subjective historic experience.

Reality is, of course, continuously changing. While some changes are structural (e.g., the invention of the printing press and electricity, industrialization and the digitalization of many processes enabled by the Internet and smartphones all have deeply changed the structure of the economy, society, and our life), others are more cyclical. For example, economies go through recurring cycles of expansion and recession.

If we examine to what extent an algorithm can deal with and even anticipate such changes to the real-life context of the decisions it makes, we sometimes find that the algorithm missed out on an opportunity to learn about the impact of such changes to outcomes either because the sample omitted certain types of instances (i.e., "rows" were missing in our data table) or because the predictors (i.e., "columns" in our data table) were inadequate.

In general, the sample should contain all kinds of environments that could be relevant for the algorithm in the future. For cyclical phenomena such as business cycles, the sample therefore should in particular cover both expansive and recessive periods. This often means that the sample should cover a rather long time period. For example, if you assume (or better, based on historic GDP growth data, establish as a fact) that the typical business cycle is seven years long, you want to consider at least seven years of data for training the algorithm (and ideally a lot more than that, as everything else being equal, a model would be more stable if it has seen two or three full business cycles).

Why does this matter? We often observe that certain relationships vary depending on where we are in a particular cycle. For example, banks have observed that a company's liquidity level is particularly important during recessions as a predictor of default risk. When a bank analyzes a company's balance sheet, it typically calculates several ratios measuring liquidity (e.g., the *quick ratio* is obtained by adding up cash and other assets that will or could become cash within less than a year (e.g., short-term investments and current receivables) and then dividing this number by the total current liabilities of the company).

If an algorithm was developed exclusively during an expansive period (with high growth, falling default rates, and banks typically aggressively trying to increase lending), it would assign relatively low weights to these liquidity ratios—in fact, some of the liquidity ratios may even become insignificant and drop out of the algorithm. Such an algorithm would be severely biased and inadequate during a recession when revenues fall and banks often restrict their lending, as both effects can lead to liquidity crunches in many companies. An algorithm developed on recession data would assign a high weight to liquidity ratios in general, and ratios insignificant during an expansion very well would now be highly significant and important factors in the algorithm.

By contrast, if the sample contained both recession and expansion data, the algorithm not only would learn that many liquidity ratios can be important but it also may figure out *when*—for example, it may increase the weight of or "switch on" certain ratios depending on the GDP growth observed in the most recent quarter.

But what happens if there is a structural break and in the future the algorithm will be confronted with situations that have never occurred in the past and therefore can never be included in the development sample? Well, it depends. Specifically, it depends on whether the past at least allows the algorithm to learn relevant (i.e., directionally correct) rules of how to adjust estimates for changes in the environment.

By analogy, consider an electronic calculator. It is possible that nobody ever has taken the square root of 8,368,703,007,682,335,001. Nevertheless, it is possible to program a general rule in the calculator to correctly calculate the square root of this number (e.g., the Newton-Raphson method). By contrast, prior to the invention of negative numbers, people had no such concept, and as a result would not have been able to program a calculator for handling negative numbers even if calculators had been around during the Stone Age.

Many changes in the real world are incremental. For example, a car insurer constantly faces new car models. When the first owners of the new model ask for a quote on a car insurance policy, the insurer has no historical data to estimate how accident-prone that new model is or how expensive typical repairs would be. This throws off insurers who price insurance policies based on historical data on a specific car model—for a new model, they need to pull the quote literally out of thin air, or they judgmentally price it based on a "similar" car model that has been around for a while.

A thoughtful data scientist can overcome this problem, however, by collecting engineering and other attributes of each car model that *explain* the loss profile. For example, when I built a pricing model for car insurance in an emerging market, I sat down with a group of mechanics to understand what drives repair cost in that country—and learned that with cheap labor, the biggest driver of repair cost was whether the car manufacturer provided locally produced (and hence cheap) spare parts or if spare parts had to be imported (and hence were prohibitively expensive). Similarly, I sat down with car engineers to identify a couple of engineering variables that influenced how accident-prone the car was (e.g., the position of certain particularly heavy parts of the car affects the likelihood of the car toppling over).

By adding such engineering variables and a flag indicating the availability of domestically produced spare parts to my sample, I could develop an algorithm that could reasonably price insurance even for a car that had never been sold in the country before. By contrast, the algorithms of competing insurers that only used historical loss rates of specific car models as predictive variables were biased (i.e., systematically off) if the design of a new car model was structurally different from previous models of the same brand or if the manufacturer introduced a premium model where spare parts had to be imported while historically it had only sold economy cars where locally produced spare parts were available.

Of course, this approach is limited—if a totally new technology emerges, all bets are off. I did not have any electric cars in my sample, hence my algorithm was helplessly biased towards combustion engines, and I doubt that it produces very accurate predictions for electric cars. It is important to recognize that certain structural breaks render an algorithm so biased to the past that it becomes unusable.

"I Told You So"

A fundamental challenge in dealing with structural breaks is that algorithms refuse to learn what isn't in the data. Imagine you are a particularly gifted data scientist and suspected a year before the global financial crisis of 2007-2008 that falling house prices would cause mortgage default rates to explode. Naturally, you would have attempted to expand the sample period to include a period with falling house prices. However, the bank you worked for only had data going back to the year 2000—and during the period from 2000 to 2006, house prices knew only one direction: going up.

This is a real dilemma. If you had added the variable "most recent change in house prices" into your model without any data from periods with falling house prices in the sample, that variable probably would have been resoundingly rejected by the algorithm as insignificant. And even if the variable had made it into your equation, you would reasonably expect that there were some non-linearities and that providing data on years with different (positive) growth rates in house prices would not teach the algorithm correctly what to predict for falling house prices.

What could you do? You would find yourself between a rock and various hard places:

- You could go back further in history. Both the early 1980s and the 1990s saw falling house prices in the US. However, even if some data on mortgages outstanding during those years could have been retrieved from some dusty paper files in your employer's basement, many of the attributes available in more recent years and featuring prominently in your algorithm would be unavailable and hence missing. This systematically missing data could seriously throw off your algorithm.

- Desperate, you also could consider creating synthetic data. You could start by copy-pasting certain instances from your sample but then manually overwriting some of the values to fabricate data from a hypothetical period with falling house prices. This not only would be a highly speculative approach heavily colored by your own biases but you would also struggle to adjust other attributes (e.g., credit card delinquencies reported in the home owner's credit bureau record) correctly; as a result, your algorithm would probably have substantial flaws and biases.

In other words, there is no good solution to expand a sample to include anticipated scenarios that in the available history simply have not occurred.

Response Speed

If historical data used for developing the algorithm can't help to anticipate future cyclical or structural changes, machine learning has one more ace up its sleeve: it can learn rapidly. Let's assume that two new books were published yesterday: this book and a 500-page epic about 50 generations of mice sharing the Palace of Versailles with Louis XIV. Your favorite online bookseller's recommendation engine now needs to decide which of the two books to recommend to buyers of Marcella Hazan's *The Classic Italian Cookbook* (a bestseller since 1976). How could it know? Given that both books are unprecedented literary achievements unlike anything that has been written before, it is impossible to predict which book will be more likely to be bought.

Machine learning offers a simple solution: it allows you to learn rapidly. Today, the first day that both books are available, you can simply display recommendations for either book randomly, collecting data on which customers click the recommendation and ultimately buy it. Machine learning is fast enough to update the recommendation algorithm overnight, so starting tomorrow the recommendation engine will correctly predict that almost all readers prefer my book (with the exception of the three people who *really* have a thing for mice and have already purchased 30+ books involving mice).

This ability to learn quickly is a major innovation of algorithms and was enabled by the advent of machine learning, flexible IT systems, and digital channels that can record user behavior in real time. It can greatly reduce stability biases because at the moment a structural break occurs, algorithms can start learning how outcomes change and thus adjust predictions.

Unfortunately, this ability to update and thus debias algorithms quickly hinges on one condition: responses must become evident quickly. Recommendation engines for websites enjoy an extremely fast response: if a webpage displays a recommendation now, it will become evident whether you click or not in the next few seconds. Even in more traditional marketing where physical letters were sent out via an ancient method called the postal service, as a rule of thumb purchase responses could be expected within 2-4 weeks.

Other situations, however, suffer from very long response rates. For example, if you develop a credit application score, depending on the product (e.g., credit card vs. personal loan) and market, it typically takes 9 to 18 months to see the bulk of defaults. This means that there will be a minimum lag of 9-18 months until your algorithm learns about structural breaks (e.g., the sudden occurrence of falling house prices). For some mortgages, we even see defaults spiking only 3-5 years after origination (this often is caused by certain tax benefits expiring then, hence changes in tax regimes can materially alter default patterns on mortgages). And when you want to explore how specific food items in a baby's diet might affect the propensity of being struck by Alzheimer's in old age, you have to wait 60 to 90 years for your response variable to materialize.

This means that algorithms predicting outcomes of processes with a slow response speed cannot self-correct easily and therefore are particularly prone to suffer from stability bias and therefore pose a particular challenge to manage and prevent biases.

Stability Bias in Defining "Good"

In the discussion so far, we looked how the relationships between predictive variables (e.g., a set of liquidity ratios derived from a company's balance sheet) and the outcome (e.g., the good/bad indicator that marks a company's default on its bank debt) change over time. Another aspect in any predictive problem that may change over time is the definitions of "good" and "bad."

Humans tend to take rather holistic and fluid perspectives when defining good and bad. For example, when defining "good food," people will not only consider multiple dimensions such as taste, texture, temperature, and presentation, they will also update the kind of factors they consider continuously (e.g., more recently "healthiness" might come into play when, for example, immediately after reading an article on omega-3 acids, their presence in a particular food might at least temporarily be elevated to a critical requirement as well).

Algorithms, by contrast, pick one precisely defined objective. For example, a search optimization algorithm might define as "good" that the user clicked a particular link and stick to it—even if to a human, it's painfully clear that the algorithm's definition of "good" is inadequate (e.g., if a particular link causes many users to click it but most immediately navigate back, it is quite clear that this link does not take users where they want to go).

When such an algorithm was originally designed, it might have been that the data scientist evaluated multiple potential metrics and found that the click-through rate highly was correlated with (i.e., a good proxy for) more difficult-to-measure outcomes such as how long a user engages with something or the user's subjective assessment of usefulness of the linked item. However, once the algorithm started directing user flows, it is possible that the world changed. For example, marketing folks might have figured out how to get the algorithm to promote their links, and spammers might have figured out how to game the system with misleading titles of their pages. As a result, click-throughs became an increasingly poor proxy for quality content—but in a classic case of stability bias, the algorithm will continue to optimize for click rates until a data scientist revisits the objective function (i.e., the good/bad definition) of the algorithm itself.

Summary

In this chapter, you reviewed the inherent stability bias of algorithms. The most important take-aways are:

- Stability bias is inherent to how algorithms work. They are designed to develop rules about relationships between observed attributes of the world around us and a specific outcome based on history, and apply these rules to the future.

- Algorithms therefore can be thrown off (and produce a harmful bias) if the world around us either changes cyclically or structurally.

- *Cyclical* variations cause an algorithm to be biased if the development sample covered an insufficiently short time span (and in particular did not cover at least 1-2 full cycles).

- *Structural* changes of the world can render an algorithm useless if the old rules simply don't apply any more but can be internalized in the algorithm if the changes are more gradual and can be tied to underlying factors (metadata) that maintain stable relationships with outcomes.

- Whenever a future situation is anticipated for which no comparable historical data exists, the mechanics of statistical model development make it extremely difficult—and practically speaking, often impossible—to embed this foresight into an algorithm.

- Rapid updating of algorithms (which, in particular, can be achieved through machine learning) is an effective way for algorithms to overcome their stability biases but this only is possible if responses to decisions taken realize with rapid speed as well (i.e., outcomes can be observed within seconds or days rather than months and years).

- Where the outcome predicted by the algorithm is only a proxy for a more holistic and complex concept (e.g., some higher-level definition of "good" and "bad"), the relationship between that particular proxy and the ultimate objective of the algorithm's users can change; as a result, algorithms also can suffer from a stability bias in their definition of good and bad.

So far, we have traced back all algorithmic biases discussed to either limitations in the data given to the algorithm or other choices made by the data scientist. In the next chapter, you will discover a naughtier side of algorithms: new biases that algorithms create themselves, out of the blue.

Biases Introduced by the Algorithm Itself

In the discussion so far, you have experienced algorithms as neutral and fact-driven and actively pursuing the goal of debiasing decision-making. The types of algorithmic biases reviewed all originated outside of the algorithm, such as in real-world biases or inadequate data. In this chapter, we will dive deeper in how an algorithm works and discover situations in which an algorithm "randomly" introduces new biases in the sense of prejudice against specific profiles of instances. Much of this can be considered noise, but every once in a while, such an algorithmic bias can be magnified, reinforced, or even whipped up by the context of how the algorithm is used, in which case the effect of such an algorithmic bias might grow out of proportion.

We first will take a deeper look at algorithmic errors and how sample sizes and case frequencies affect the working of an algorithm. We then will dive even deeper in how these issues can play out in a particularly harsh way in

© Tobias Baer 2019
T. Baer, *Understand, Manage, and Prevent Algorithmic Bias,*
https://doi.org/10.1007/978-1-4842-4885-0_10

tree-based modeling approaches, which are behind many machine learning models. Finally, we will discuss the implications of all of this from a more philosophical perspective.

Algorithmic Error

Humans are used to making binary predictions, such as "It will rain today!" or "Don't hire her; I don't think she's up to the task." or "Let's buy that one; it looks yummy!" Algorithms, by contrast, live in a probabilistic world: they say "83% chance of rain," "12% chance of this candidate being rated good or better 12 months into her job," or "98% chance of you liking this cake." In fact, even though many algorithms are designed to predict a binary outcome (Will it rain today, yes or no?), the way they are constructed many typical algorithms are mathematically incapable of assigning a probability of zero or 1 to any event; it's only between 0.000......1% (and possibly a lot more zeros) and 99.999% (and possibly a lot more 9s).

When algorithms are used to make decisions, there is always a decision rule stating "Do this if the estimated probability is larger than x" or "smaller than y." Every yes/no decision therefore always implies an algorithm saying "by the way, there is this non-zero probability that my prediction is wrong; that it won't rain, this college drop-out applicant will be a superstar, and you'll hate that chocolate sponge cake with chili and bacon."

This implies an interesting conundrum: like a lawyer coating all his statements in clever disclaimers, we never can point our finger at a particular prediction and accuse the algorithm of getting it wrong; the algorithm always will lift its finger and coldly point out that it indicated exactly the probability of its prediction being wrong. In fact, the algorithm's default approach to assessing a situation where it has exactly zero insight is to assign the population bad rate. In other words, if the data you have shows that in your location there are on average 183 days of rain per year, a clueless algorithm will deftly indicate a 50% chance of rain.

The only way to demonstrate that an algorithm has made an error is therefore to consider a whole group of instances and show that the number of "good" versus "bad" outcomes is significantly off from the algorithm's prediction. For example, if we take 1,000 call center agents that the algorithm predicted had less than 10% chance of hitting the minimum performance level but that we hired nevertheless, maybe because we accidentally misread the algorithm's output (True story: I have several clients who stumbled over an algorithm's performance issue because they accidentally did the opposite of what the algorithm advised and noticed the error only later!), and if we find that 870 of them (that's 87% of 1,000 agents!) not only hit the minimum performance target but as a group outperformed 75% of the other call center agents, then we can argue that the algorithm is significantly off and hence rubbish.

Establishing whether an algorithm's outputs are correct or not therefore is a number's game, and that leads us to the issue of sample size and case frequency, which we will tackle in the next section.

The Role of Sample Size and Case Frequency

At the heart of model development is the tension between overfitting (reading rules into random patterns) and lack of granularity (omitting variations of outcomes for specific subsegments). Let's explore this issue with an example: a search ad optimization algorithm tasked with estimating how likely a particular user is to click an ad for walking canes. More specifically, let's assume that the owner of the webpage where the ads would be displayed knows the user's age, and you therefore now are concerned with capturing age as a predictive variable.

You intuitively know that kids rarely use walking canes and that therefore there must be some age limit above which the use of walking canes really takes off. You also may hypothesize that at some point folks may have such serious ailments that they are more likely to need a rollator or cannot walk at all any more, and that therefore walking cane use may ebb down again from a certain age onward.

If you have data on the entire United States population, you have more than a million data points for each year of age from 0 to at least 80 years of age.[1] You can not only calculate for each year of age the likelihood of clicking ads for walking canes but have enough data to split it further to examine men and women separately, or if you believe in astrology, to compare the propensity of Pisces to use walking canes with that of Aquarius. You can even cut the data by age in days (that's 365.25 times more granularity than by year) in order to reveal if people are more prone to look for walking canes on the day before their birthday.

Unfortunately, you very often have a lot less data, however. Assume you have a small sample with just three people who are 95 years of age. None of them uses a cane. Does this mean that 95-year-olds have a zero chance of using walking canes? No, very possibly these three folks are exceptions. You really can't say whether out of a thousand 95-year-olds, none, 100, or maybe 997 would have a walking cane. In other words, you have encountered a situation where the sample size is too small to make a statistically significant inference. This is the daily fate of data scientists trying to develop good algorithms. So what can you do?

---

[1]U.S. Census Bureau (2011), "Age and Sex Composition: 2010," 2010 Census Briefs, May, 2011.

The default solution of an algorithm is to combine your merry three 95-year-olds with other but similar cases. Maybe there are two 96-year-olds, five 94-year olds, and a whole bunch of people 90-93 years old. An algorithm therefore might create a group of people that are 90+ years old or maybe 90-99 years old. How big a group do you need to make a statistically significant inference? This depends on how frequent a phenomenon you examine. If 20% of people in this age group use a walking cane, statistical methods (here we use the adjusted Wald technique) suggest that with 95% confidence you can expect to find between 13 and 29 walking cane users in a sample of 100, so a small sample is sufficient to give you a rough idea at least (but the propensity you measure might be off by 5-10 percentage points). If you are investigating a rare disease striking only 1 in 10,000, your chances of finding even just 1 case in a sample of 100 are, of course, abysmally small.

What does this mean for your algorithm? The chances for your algorithm to be able to correctly assess a particular type of people (e.g., 95-year-olds) are higher the larger the sample, the closer the average propensity of the binary outcome to be one or the other to 50%, and the more frequent this particular type of people occurs (even if you have data on the entire US population, you may have just 1 or 2 people who are 114 years old). The further you are away from the ideal situation (i.e., the smaller the sample, the more imbalanced the ratio between good and bad or yes and no outcomes in your sample, and the more infrequent the particular type of instance you consider), the greater the pressure on your algorithm to throw this case in a big pot with many other instances that are different.

This aggregation problem is particularly thorny for categorical variables. Combining 95-year-olds and 96-year-olds appears defendable, but what about a job category of "others?" Imagine someone has developed some bacteria that produces oil, and he sells it in small batches on the Internet to campers who want to be carbon neutral. Is he a farmer or an entrepreneur in the energy business or maybe more of a retail trader given that marketing, packaging, and shipping constitute most of his cost?

The implication is that infrequent, atypical cases are likely to be lumped together with other often similarly infrequent and atypical cases that may or may not at all be similar. As a result, the model develops a bias in the sense that the behavior of a probably irrelevant reference group is projected on such rare cases. For example, if a bank's credit scoring model treated our ecological camping oil manufacturer as a farmer, and if in that market farmers are generally low risk because of government subsidies and price controls, it may seriously underestimate the risk of this novel niche business.

The Issue of Tree-Based Models

There is a range of structures for algorithms that naturally all come with their own advantages and disadvantages. In Chapter 3, you saw an example of a linear model. A strength of this model structure is its efficient use of the data available: for each input factor x, the data of the entire sample is used to estimate the parameter β, and all input factors are considered in estimating the outcome for a given instance. As a result, such models often still can do justice relatively well if they encounter rare profiles. In the example of a bank evaluating a loan application from our bacteria-derived camping oil seller, the scorecard may combine three inputs: the risk index derived from the industry, a liquidity ratio, and the age of the business. Even if the industry risk is underestimated, the scorecard still may penalize the company for its scant liquidity and lack of operating history.

There is a different class of modeling approaches that uses data in a much more wasteful manner: tree-based approaches such as decision trees and random forests (ensembles of possibly a hundred or more decision trees). These approaches are becoming increasingly relevant because they are a lot more flexible than linear combinations of features and at the heart of many machine learning techniques.

Trees are a classification approach that divides up a population into smaller and smaller parcels such that each parcel in the end is intended to be a quite homogenous group for which the average outcome observed in the development sample is representative. In our credit scoring example, the first factor the tree considers might be the industry. It may create three broad groups of companies based on trees: agriculture, manufacturing, and "other" (such as services). For manufacturing, it may in a second step divide companies into high and low liquidity companies; in a third step, the low liquidity companies may be split by their length of operating history.

For agriculture, however, the tree might follow a different logic. For example, on the second level, it may split agricultural operations that own their land and real estate from those that lease it; the companies that own their land may then further be split into crop-producing, livestock-producing, and "other" businesses (that may combine odd fellows such as our bacteria-based oil producer and producers of decorative fish).

The issue is that the sample literally gets split up. If you start with 10,000 companies in your sample, maybe 800 are in the agricultural sector. Of these, 500 own their land, including your bacteria-based oil manufacturer. These 500 then get split into 230 crop producers, 240 livestock producers, and just 30 companies in the "other" criteria. Such a small group already stretches the algorithm's ability to calculate a meaningful default rate (in fact, by all rules of thumb the sample size would be insufficient, given that for many small and medium enterprise portfolios bad rates are in the range of 2-5%). And it will be all but impossible to split it further by any other criteria such as a liquidity ratio or the age of the company.

In other words, what happens here is the following:

- Our company is lumped together with a small group of other unusual companies that may or may not be at all comparable to our company.

- Because this group has only a very small number of companies in it and the tree-based modeling approach is so wasteful and hungry for data points, the tree doesn't have enough data to further differentiate this group by certain other factors that are predictive of risk such as liquidity and age of the business.

- Our company is assigned the average default rate of its group, which may or may not be a reasonable estimate of that specific company's risk.

Now, it is not true that decision trees are completely random. The tree above did try to make use of metadata (e.g., the data scientist provided the information that our company could be considered to be an agricultural enterprise), and the groupings reflect the dependent variable, so if our bacteria-based oil producer did default in the end, the algorithm would tend to group it with other high-risk companies, while if the performance was good, the tree would try to group it with low-risk companies. However, a single such company really doesn't allow any generalizable insight on the typical risk of such companies, and to the extent that outcomes informed the grouping of rare cases, tree-based models are prone to a type of misclassification called *over-fitting*. If such rare instances are grouped together simply because they had the same outcome (e.g., they all defaulted or did not default), the model will perform really well on the sample but is likely to be terrible at predicting outcomes in the future because there is no underlying robust logic.

A Philosophical Perspective

This chapter talked primarily about the misclassification of rare cases where the amount of data available is simply insufficient to make any statistically valid inferences on the nature of this type of incidents. As a result, the algorithm is "biased" because it either projects the performance of a more or less irrelevant reference class onto such an incident, or it generalizes the outcome of these rare cases (i.e., the value of the dependent variable) to all similar cases.

On the one hand, one can argue that this is not a big issue:

- These cases are, by definition, rare exceptions. If there were more such cases, the heft of the data would nudge the algorithm to correctly estimate outcomes for this type of situation (i.e., without bias).

- It is not even straightforward to "prove" that the algorithm is wrong. The algorithm does implicitly assign a non-zero possibility to being wrong, and without additional data (expanding the sample size), there typically is by definition insufficient data to "prove" that the estimated probability is significantly different from the algorithm's estimate.

On the other hand, one also can raise two serious issues with this:

- When an algorithm assigns a certain estimate (e.g., an 83% recidivism rate for a given felon), it doesn't indicate the degree of uncertainty associated with this estimate. It might be that there were thousands of almost identical cases in the sample and the prediction was made with a 99% confidence interval of +/- 1%, so if you had 1,000 of such felons, the algorithm would bet its life that between 820 and 840 of them (that's what 83% +/- 1% means) will recidivate, or it might be that the felon in question was a very odd case where the 83% estimate is the random result the algorithm obtained by lumping together a handful of similarly odd fellows. As a result, one could argue that an overall highly predictive algorithm acts a bit deceitfully if it assigns estimates for cases where a better answer might have been "I don't know."

- A random prediction by the algorithm could become self-fulfilling if it influences decisions that impede the generation of contradictory data points. For example, if an algorithm deciding on graduate student admissions has randomly developed a bias against female engineers and it therefore prevents any female applicants from pursuing a degree in engineering, there will never be any data to prove the algorithm wrong.

In my opinion, the gravity of the issue is an ethical question that depends on the purpose of the algorithm. An algorithm that picks personalized elevator music for your daily vertical commute in your outlandish apartment building is unlikely to cause much harm if it develops an unreasonable bias against Johnny Cash; an algorithm driving decisions that have material impact on people's lives can cause serious harm. We will examine this question of the gravity of a specific bias in greater detail in Chapter 13; for now, keep in mind that misclassifications are in the nature of every algorithm but that for unusual incidents there is a greater risk of estimates that are completely off, and that this may cause serious harm for the affected incidents.

Summary

This chapter dealt with algorithmic biases that develop randomly if the sample is insufficient to train the algorithm to adequately assess certain rare incident profiles. You in particular learned that:

- Every estimate of an algorithm comes with a certain probability of error.

- While for typical cases (where there are many similar incidents in the sample) algorithms by design are correct at least on average, for rare (unusual) cases there is a particular risk of a completely wrong estimate that is not even correct on average.

- This happens because the algorithm either lumps unusual incidents together with other cases that may be a completely irrelevant reference group, or it extrapolates outcomes from a very small and therefore statistically insignificant number of incidents (so-called overfitting).

- Such biases are more likely to happen for small sample sizes, and more precisely, if the sample is small relative to the diversity in the sample. A sample of 50,000 can be considered large if there are only a few dozen different types of incidents but it's tiny if there are millions of different types.

- The issue of sample size is aggravated if the modeling technique is data-hungry, such as tree-based approaches that make very inefficient use of data.

- While the issue by definition will affect only rare types of incidents, this type of bias still should worry users if the impact of the bias is severe or if the affected type of incident was rare only in the sample, but is more common in the general population to which the algorithm is applied.

All biases reviewed so far were static in the sense that a data scientist designed a model, collected data, and estimated a model that then exhibited certain biases. In the last chapter of this part of the book, we finally will dive into the particularly challenging world of social media where biases develop dynamically through the interaction between algorithms and users.

Algorithmic Biases and Social Media

This final chapter of Part II of the book is a bit different from the previous chapters insofar as it zooms into one particular situation where algorithmic bias occurs: the choice of posts shown to social media users. In doing so, I achieve two objectives: this serves as a case study that shows how the various biases discussed so far can interact and reinforce each other, and it illustrates how algorithmic bias can be dynamic. Rather than set in stone, the bias can develop and grow over time out of an interaction between user and algorithm—the biases of both reinforce each other until the result can be a discomfortingly strong distortion.

I will approach the topic in three steps. First, I will show how algorithms shape the flow of news or posts on social media and how algorithmic bias can affect this flow. Second, I will discuss how behavioral biases of the user shape his or her news consumption and through this the algorithm choosing news and posts. Finally, I will derive implications for combatting algorithmic bias in social media (and in doing so also start moving towards the question of how to counteract algorithmic biases, which is the focus of the remaining two parts of this book).

© Tobias Baer 2019
T. Baer, *Understand, Manage, and Prevent Algorithmic Bias,*
https://doi.org/10.1007/978-1-4842-4885-0_11

The Role of Algorithms

Chapters 4 and 7 covered how much model design can drive algorithmic biases. The case of social media in particular highlights the importance of the definition of the outcome variable to be predicted and the choice of explanatory features.

Algorithms choosing items to be displayed in user interactions such as a social media news flow often rank by the likelihood of an item being clicked. This model design choice has interesting implications:

- Strictly speaking, it means that rankings are not about the actual content of the news item at all: they are strictly about the headline or the accompanying picture or whatever else is displayed.

- They therefore model in particular to what extent an item will attract the user's attention (remember how our brain filters out most things we see and hear) and evoke a desire to explore the item further.

For example, if an algorithm has "understood" that you tend to click items mentioning cats, an article titled "Cat and mouse games in Rotterdam" would probably get a high rank and show up in your news feed. Sure enough, your brain is likely to alert you of an item involving "cat," and maybe your brain's reward system would weigh in with a believe that an article about "games" must be fun and hence a pleasant and rewarding read. As soon as you click it, however, you will realize that the article is really about customs investigators in the port of Rotterdam trying to stay abreast of constantly involving tactics of smugglers; within a split-second you will have clicked the back button.

Alternative model design choices would have resulted in very different outcomes. Predicting the time you spend on reading a piece clearly would have banned this article to the bottom of the stack (pulling up instead an article titled "Furry customs" because, in fact, it deals with cute rituals involving pets in general and cats in particular). A particularly high-browed algorithm even might try to predict how much you have "learned" from an article—it may be a while until we are wearing brain scanning devices giving the kind of real-time feedback required to model this, but in the meanwhile, predicting actions such as "liking," forwarding, or saving an item could be useful proxies.

Another important aspect of model design is the choice of predictive factors. With thousands of words in every language, a simple flag for the occurrence of each word (e.g., "cat") would make a very unwieldy algorithm running into all kind of statistical problems (around significance). Instead, data scientists create aggregated features that crystalize important characteristics, such as "mentions an animal" or "contains an emotionally evocative word that was mentioned in at least 20% of the last 50 items the user clicked."

The design of these aggregate features therefore can play an important role in the development of algorithmic bias. For example, a data scientist convinced that the political bent of a news item is a key driver of clicking behavior might develop a complex semantic metric measuring the right vs. left bent of an article on a scale from -1 to +1. Now imagine that the overall algorithm is poorly designed with few other features offering any real insight into the user's choices. What would be the consequence? The algorithm's best bet to rank items would be the political bent of an item; if historically the user clicked somewhat more often on articles leaning to the left (maybe because one of the leftist parliamentarians has the last name of "Cat" or a sloppy semantic metric considers the name "Catlin" an incidence of a cat), the algorithm now would prioritize strongly left-leaning articles (and maybe suggest that the user becomes a follower of Mr. Cat's social media feed, too).

Again, the data scientist has choices. A metric of an item's "fun factor" is likely to predict both click-through and engagement but may discriminate against well researched but dry articles rich in hard facts; in fact, the "fun factor" algorithm might really struggle to differentiate satire from the most outrageous fake news and serve up so many of the latter that the user starts to mistake even the former as "facts."

Let's explore a bit deeper how the data scientist's choices can influence the shape and severity of this modeling problem. For example, the data scientist could create metrics of the extent to which an item is what the user expects, or metrics of the degree to which a news item will change the customer's perspective, or metrics of the degree to which a user has been underexposed to a particular perspective. How would this affect the slant of the algorithm?

Aligning the algorithm with the user's expectations clearly would feed the user's confirmation bias—you don't need a PhD in Psychology to see what feeding a Zeta Reticulan articles only about Martians assaulting Zeta Reticulans leads to. Optimizing the algorithm to achieve maximum changes in the user's beliefs might look like a great way to identify the most "informative" and "useful" news but actually would be a particularly manipulative approach—the algorithm would aim to change a user's perspective for the sake of change. In fact, if we believe that the starting point of views of a population is a distribution where most users have moderate perspectives, by definition this algorithm would aim at polarizing perspectives by picking the most spectacular and scary fake news because the only way you can change a moderate view is to make it more extreme in one direction or the other!

Measuring the extent to which a user has been underexposed to particular perspectives per se seems useful to render users better informed and help them to break out of "echo chambers"—however, given the unlimited number of topics (I can think of thousands of topics I've been underexposed to, ranging from the design of musical instruments in the 12th century to the chemical engineering of air refreshers), the data scientist will have to decide exposure to which topics to measure and therefore again would massively influence the direction of the user's enlightenment. Just imagine the health and safety problems a data scientist might be exposed to if she gets known to have willfully excluded climate change from her list of important topics!

The dilemma is that, as Richard Thaler points out in his book *Nudge: Improving Decisions about Health, Wealth, and Happiness* (Penguin, 2009), for many similar situations in life, the data scientist selecting items for a social media feed is a choice architect and therefore does not have the option to be "neutral." And even well-meaning choice architectures can backfire. For example, it would be plausible to argue that "easy to understand" news items are more useful; however, brevity will be an excellent proxy for ease of understanding and again bias the algorithm against well-researched articles backing up statements with lots of figures. Similarly, an indicator of the journalistic robustness of an item sounds like a clever feature (e.g., a rather intelligent algorithm might be able to differentiate an attention-seeking discussion of a single case short on facts with a rigorously researched piece offering tons of facts and figures citing numerous reliable sources) but may discriminate against philosophical pieces pointing out important conceptual issues with core beliefs of society as a whole (e.g., when Nassim Nicholas Taleb chastens economists for disregarding how scale exponentially increases complexity in his book *Skin in the Game* (Random House, 2018), he is short on numbers but long on insight and implications).

One final issue in the choice of data for modeling click-through behavior shall not be missed: there is a potentially fatal feedback loop if an algorithm is repeatedly refined based on click-through behavior on links selected and sorted based on earlier versions of the same algorithm.

When a user peruses a news feed or any other list of items, there normally is a default directionality: in all major languages, you go through a vertical list from top to bottom. If items are arranged in a horizontal bar next to each other, an Arab speaker would naturally go from right to left while an English speaker would go from left to right.

This creates an interesting artifact: given that every user will browse a limited number of items (e.g., until fatigue sets in, the phone rings, or the cat (the pet, not the socialist politician) jumps on the keyboard), statistically in an otherwise randomly ordered list of items, the first item is most likely to be clicked, while for each position down the likelihood of being clicked will drop. This effect is reinforced by a second psychological effect: users often expect items to be sorted, so they therefore subconsciously assume that the first item is particularly important or relevant.

The consequence is that if a hapless algorithm starts prioritizing news items mentioning "cat" out of some sheer fluke and puts them at the top of the items in the news feed, the data in due course will show an uptick in clicks of anything feline—and the initial algorithmic fluke will become a self-fulfilling prophecy and a new algorithmic bias has been born.

The Role of Users

This feedback loop points us to the role of the user in algorithmic bias. In fact, there are three types of users—the consumers of social media (i.e., the users reading them), those individuals who use social media as an outlet to distribute their views and articles, and the owners and managers of social media assets who try to earn money by facilitating the distribution of content from producers to its consumers. All three can and do contribute to algorithmic bias.

Users expose, of course, the full gamut of biases reviewed in Chapter 2. The way a user browses social media is not unlike how our ancestors might have strolled through nature on the quest for food. Here the bright red of some berries might beckon; there fresh dung or an animal's imprint in mud might promise prey. Naturally, therefore, action-oriented biases will guide the user's split-second choices of where to click.

One action-oriented bias particularly relevant for social media is the *bizarreness* effect. My own LinkedIn blog illustrates the point: I found descriptive titles dryly stating the subject (e.g., "How to solve the small data dilemma in wholesale credit risk modeling") garnered fewer clicks than puzzling ones (e.g., "What the $450 million Da Vinci auction means for credit application fraud")—you yourself might find the second topic a lot more interesting!

Amongst the stability biases, *anchoring* can be blamed for prodding the user to click the first item in the list—merely by being the first item on the list, it anchors the user's perception of what is important and *en vogue*.

Also, pattern-recognition biases affect the user. The abundance of items on social media challenges the brain's capability to synthesize an excessive amount of information. The *Texas Sharpshooter bias* and *stereotyping* mean that a couple of headlines (e.g., on the behavior of Martians) could easily inform broad prejudices. Unfortunately, the *confirmation bias* filters items for conformity with existing beliefs, literally drilling fledgling prejudices deeper and deeper into the mind.

Interest biases go hand in hand with these cognitive biases. Browsing social media is a recreational activity—the user's objective is often relaxation and entertainment. The consumption of a news item will either evoke a pleasurable emotion (e.g., joy and laughter, or at least the satisfaction of seeing one's beliefs confirmed) or a painful emotion (e.g., fear triggered by reading about a threat, or the pain of cognitive dissonance if a news item disputes cherished beliefs). If a user's objective is to relax and be entertained, the brain naturally will prioritize items expected to trigger a reward reaction in the brain—just how our ancient hunter-gatherer ancestors experienced a spike in their brains' reward systems every time they discovered a new food source. Similarly, a brain in the pain state or a neurotic personality would be drawn towards items informing about dangers or confirming fears.

In fact, interest biases breed confirmation biases at a scary level. Jeremy Frimer of the University of Winnipeg demonstrated that individuals who are paid for the engagement with arguments that either support or challenge a belief with high emotional valence (in his experiment, he used gay marriage) are literally giving up money for not having to engage with challenging arguments.[1]

This experiment puts a price tag on the cost of what psychologists call *cognitive dissonance*. Once an individual has formed an opinion on something, the brain "locks in" its decision and actively opposes efforts to change the decision. This phenomenon bears the handwriting of nature's energy efficiency: given that exploring and evaluating options burns energy, it is wasteful to reconsider the same question again and again unless the rewards of doing so merit the effort. The brain seems to accomplish this through cognitive dissonance—if we have made a decision but are confronted with a piece of contradictory evidence, we literally feel pain, which is nature's way of saying, "Don't go there unless the pain of going somewhere else is even higher!" This is where confirmation bias is at its strongest.

Social biases join the fray, too. For example, the desire to conform with group norms might bias users towards items informing about the opinion or actions of important people (which is why a headline starting with "President says ..." might get more clicks than "Tobias Baer says ...," in spite of all my brilliance!).

The user's actions are not only biased on the "micro-level" of choosing which news items to dwell on, however; they are also biased at the "macro-level," where the user chooses which sites to frequent and whose news flow to subscribe to. If you're an avid member of the Martian party, you naturally would subscribe to their updates rather than the updates of the rival Zeta Reticulans' party. The result is the creation of so-called "echo chambers" where social media creates virtual communities of individuals who share a particular, often very narrow perspective on an issue and produce for each other a news flow that perpetuates their perspectives while starving each other of contrarian views.

---

[1] J.A. Frimer, L.J. Skitka, and M. Motyl, "Liberals and conservatives are similarly motivated to avoid exposure to one another's opinions," *Journal of Experimental Social Psychology*, 72, 1-12, 2017.

If we observe a user's behavior over time, we can note another important aspect of bias. When someone has to make a decision (e.g., which shampoo to buy, which party to vote for, whom to marry), there is an exploration phase, a decision point, and a post-decision phase.

During the exploration phase, users have a desire to understand their options—especially if the brain is in a "pleasure state" (i.e., users are relaxed, there is no time pressure, and users are not preoccupied by other worries), so confirmation biases can be actively suppressed and novelty is sought out.

As the decision-point approaches—not least because users become exhausted from their explorations—users need to evaluate their options, and as this is a demanding cognitive task, a plethora of biases will be ready to come to the rescue. Once a decision is reached, however, the brain "locks in" its decision and actively opposes efforts to change the decision, as discussed above, and the confirmation bias keeps future deliberations to a minimum, at least until a sufficiently strong impulse to challenge one's beliefs can compensate the cognitive dissonance and break the confirmation bias. If you think of acquaintances who had to deal with adultery, you might have observed both: spouses refusing to accept an ocean of tell-tale signs that they are cheated on, and explosions when the evidence of cheating was strong enough to shatter the belief in a faithful spouse.

As always, however, personality does matter, also with respect to the confirmation bias. So-called neurotic personalities are constantly scanning for dangers; to the extent that they consider a wrong decision a danger, individuals with a strong neurotic trait therefore may exhibit a strong appetite for reconsidering decisions they have made in the past (possibly driving everyone else around them mad and paralyzing themselves).

Algorithms are not the only ones that will tap into these behaviors and biases of users—there are many people out there with an interest in influencing the decisions of users (e.g., sellers of shampoo, political parties, and people who want to marry you). And they can take algorithmic biases to a new level.

While users intend to make the best decision and their brain uses biases only as a shortcut in order to save mental energy, and while algorithms at least are based on statistical principles aimed at avoiding bias, people with a strong interest in influencing a user's decisions are tempted to take actions that exploit biases for their benefit.

If you revisit the user biases sketched out above, you can probably think of ways to shape a news item or other social media post to cater to most or all of these biases. For example, headlines and accompanying pictures often are designed to trigger the *bizarreness effect*. A milder version of this is to stress the newness of something, similar to consumer good marketers often touting in big red letters that a product is NEW (or has a NEW formulation). A simple way to *anchor* an item through position is to pay a website to put one's item on top of a list. And celebrity endorsements and pointing out that an item is very popular are ways to trigger *social biases*.

But what happens if an algorithm is picking items for a social media feed? The agent trying to influence users now faces two hurdles. First, the agent needs to get the algorithm to pick the agent's item and assign it as high a rank as possible, and then the user must be enticed by it to click. And here algorithms can be used in a totally new way: not just as a way to predict an outcome but as a guide to create items (here social media postings) that will have the desired outcome.

Let's assume you knew the algorithm used by a social media website gave a news item 10 points each for each incidence of a cat in either the headline or the headline picture, 5 points for each incidence of a cat in the body of the article, and 1 point for each mentioning of another type of pet. How could you help the Dutch customs office spread an article teaching citizens how to help spot financial crime? Easy: Get as many cats into your article as possible by titling it "Mr. Cat explains how copycats let the cat out of the bag."

Of course, rather than gaming the algorithm, one also can simply pay for access to social media users and then exploit the users' behavioral biases to influence them. This strategy was exemplified by the now-defunct Cambridge Analytica, which used psychometric scores to derive from a Facebook user's "likes" a behavioral profile (using the popular OCEAN Big 5 framework that describes human personality through five dimensions).[2] Clients of Cambridge Analytica then paid Facebook to place ads (supporting Brexit or the Trump campaign) that pulled exactly the right "triggers" to influence a user based on the user's psychometric profile.[3]

This risk of manipulation is one of the reasons why many algorithms—including those picking social media posts for consumption—are kept secret. However, this is easier said than done. If you produce a large number of posts and systematically measure which ones are ranked highest, you create data that can be used to reverse-engineer the secret algorithm—you literally build a second algorithm that predicts how the secret algorithm will decide, and with enough data you can actually build a virtually perfect replica of the "secret" algorithm.

---

[2] **O**penness to experience, **C**onscientiousness, **E**xtraversion, **A**greeableness, **N**euroticism
[3] www.theguardian.com/technology/2017/may/07/the-great-british-brexit-robbery-hijacked-democracy

Consumers obviously will be annoyed if they are duped into reading content they neither expected nor appreciated, and if they find that one source of social media is particularly annoying, they will quickly flock elsewhere. And that calls in the third group of stakeholders: the owners and managers of social media assets (e.g., sites) who have an economic interest in keeping consumers glued to their service.

How can they change outcomes? They need to find ways to sabotage attempts by producers of social media to unduly influence their algorithms (e.g., by producing fake news). This may entail new algorithms, such as algorithms to assess the likelihood that an article is "doctored" (e.g., contains a suspiciously high frequency of the word "cat") or is pushed by a known perpetrator. As you can see, very quickly an algorithmic arms race will break out; most likely you have heard about this already in the related domain of search optimization (i.e., the dark art of manipulating results of Google and other search engines, such as those of online shopping sites).

Is your head swirling from this crescendo of algorithmic biases? Rather than confuse you, I merely tried to illustrate a central point: algorithmic bias is not a simple kink in a formula. It's the combination of multiple factors, creating a problem at what is called the "system level." This means that in order to fix the problem, we cannot simply flick a switch in the algorithm—we must change how the system works.

Implications

In his book *The Lucifer Effect* (Random House, 2007), Philip Zimbardo explores the question of what makes people commit horrendous acts (in his mock prison at Stanford University and in real military prisons). He finds a tendency for people (society, the press, peers) to finger individual actors as "bad apples" while at the same time ignoring how the system has made innocent individuals behave like bad apples. There is a parallel between rogue prison guards and algorithms: both can easily be fingered as the cause of the problem at the peril of ignoring the deeper root causes in the system.

Our discussion above has clearly shown how the mechanics of algorithms can create, reinforce, and perpetuate biases in social media. But we also have seen two strong forces that are working to embed biases even in the most innocent algorithm of the world: users themselves exhibit very strong biases—to the extent of even accepting monetary losses in order to avoid the pain of being exposed to bias-busting information—and other stakeholders with strong economic benefits have an interest and techniques at their disposal to create news items towards which the algorithm exhibits a bias.

The solution to the problem therefore must address the system rather than focus just on the algorithm. Here I just give a couple of examples of what this could look like:

- Mindful of the effect of rank in a list, a website could randomize the order of items displayed once an algorithm has prioritized the, say, 20 or 50 most relevant items to display. This would break the feedback loop of users tending to click the top items first and thus is an example of how a tweak in the system can generate better, less biased data.

- Aware of one's role as a choice architect, one could replace the simple objective of "let's maximize click-through rates" with a multi-tiered approach that, say, first picks top stories but then complements the list with a clearly labeled "contrarian view"—at least in the exploration stage, some consumers might actually seek out these items.

- One also could empower users to be part of the solution, rather than paternalistically trying to decide what is best for them. For example, users could choose in site settings how strictly to filter out possibly fake news or to what extent to provide opposing views by design. They also could choose to see such news items with a warning label ("Fake?") instead of the algorithm simply eliminating them.

- Willful manipulation of algorithms—especially for economic gain—can be fought not only by owners and managers of social media platforms; just as false advertisement and defamation can have legal consequences, also laws and regulators can play an important role in safeguarding the integrity of algorithms driving the social media news flow.

Summary

In this chapter, you traced algorithmic bias in a particular context—social media—and found through our inquiry how different elements in a system can collude to create algorithmic bias. In particular, you learned that:

- Statistical formulas can collude with biases of users and personal interests of individual stakeholders; often algorithmic bias is the result of such collusion (as opposed to the result of the algorithm on its own).

- Just like a perforation will predetermine the most likely line along which a piece of paper will tear when exposed to forces, model design choices made by the data scientist can determine the type of biases an algorithm is most likely to develop.

- The data scientist's dilemma in making these design choices—which sometimes are called *choice architecture* in behavioral economics—is that choices often come down to choosing between biases working in different directions while "no bias" is not a feasible option.

- If algorithms are refined over time (as it is typical in particular for machine learning algorithms), there is a risk of a feedback loop that over time amplifies initially small biases.

- Users of social media feed their biases into the algorithm during its development, essentially exercising the force that causes the perforation to tear.

- Users are exposed to a wide range of cognitive biases, including anchoring, the bizarreness effect, and social biases; however, confirmation bias can be considered the strongest bias affecting how humans use social media.

- Counteracting these biases can literally cause pain to the user (called cognitive dissonance); this means that the user's behavior can have a very strong direction that cannot easily be changed.

- On the positive side, confirmation bias arises only once a user has formed a decision; this means that during the exploration phase of a topic of interest, a user might be open to measures fighting bias.

- Interested parties—including individuals who want to steer algorithms to prioritize their posts as well as owners and managers of social media websites—have forceful ways to impact the interplay between users' behavior and an algorithm, steering algorithms both towards specific biases and away from biases.

This illustrated why fighting algorithmic bias requires actions at multiple levels. Just as reducing road accidents requires a range of actions—ranging from better car technology to training of drivers to traffic laws to physical modifications of roads—reducing algorithmic bias requires a range of actions at the technical (algorithm) level, the user level, the business (or process) owner level, and the regulatory level.

How exactly to do this is the subject of the remaining two parts of the book. In the next (third) part, we will review ways users of algorithms can tackle bias; in the final (fourth) part, we will discuss on a more technical level what levers data scientists have at their disposal.

This chapter illustrated algorithmic bias in the context of social media where the term "user" has a relatively specific and narrow definition—namely first and foremost the individual consuming social media—although users also can generate content and thus sit on the opposite side of the social media content flow. In other contexts, even the owner of an algorithm is considered a user— for example, in banking, one would typically consider the bank (as an institution) a user of algorithms (as opposed to using human judgment administered by staff). In the rest of this book, I therefore will use the word "user" in a very broad and encompassing way that includes both the people who deploy algorithms to serve a certain purpose and the people who are exposed to the outputs of algorithms in the context of a specific decision.

What to Do About Algorithmic Bias from a User Perspective

Options for Decision-Making

In the first two parts of this book, you learned that the mechanics of algorithms expose them to many potential sources of bias, that biases are real and at times exceedingly harmful, and that algorithmic biases very often originate in real-world biases.

This part of the book will discuss how users of algorithms (e.g., business leaders and government officials)—as well as others who need to make decisions about the use of algorithms, such as compliance officers and regulators—can detect, deal with, and prevent algorithmic bias. As you will see, the tools and techniques available to business users are often complementary to the contributions data scientists make in fighting algorithmic bias, which will be the focus of the fourth and last part of the book.

And as mentioned, I will use the word "user" in a very broad sense that includes both individuals or institutions who deploy algorithms to inform certain decisions and the subjects who are exposed to the outputs of algorithms, such as an individual or company requesting an approval.

Before diving head-on into biased algorithms, however, I want to take a big step back and stress that ultimately, our foe is bias per se. Focusing just on managing and eliminating algorithmic bias is inadequate in two important dimensions:

1. Algorithmic bias only can arise if there is an algorithm. Just as one effective strategy for not being robbed is to

© Tobias Baer 2019
T. Baer, *Understand, Manage, and Prevent Algorithmic Bias,*
https://doi.org/10.1007/978-1-4842-4885-0_12

not enter a dark alley in a shady neighborhood at night, sometimes we simply should not use an algorithm. This opens up a much bigger solution space for dealing with algorithmic bias.

2. At the same time, if we choose not to use an algorithm because it's biased, we still have a decision to make. We must be careful not to move from the frying pan into the fire: our alternative approaches to making a decision might be a lot worse! Therefore we need to see algorithmic bias on a relative basis, compared to other options—after all, we still might embrace a slightly biased algorithmic as the smallest of all evils.

Let's therefore first discuss the broader picture of a decision-problem where an algorithm could be deployed, and then explore a simple framework to assess whether the benefits of an algorithm might outweigh the inherent risks of algorithmic biases.

Definition of the Decision-Problem

Fundamentally, algorithms are deployed as a means to *differentiate* treatment of different people. The most basic alternative to using an algorithm is to either treat everyone the same (e.g., you could try to predict for each visitor of your website what book recommendation would be most useful, or you could make your life easier and simply recommend this book here to *every* visitor of your website) or to toss a coin (e.g., decide which candidate to hire through a draw). This is not the norm but it is also not as farfetched as it sounds if the algorithm is highly biased: I had several clients with a credit scoring system that performed *worse* than a random number—in all four cases the algorithms had been blindsided because the data used to develop them or the inputs fed into the algorithms had been heavily biased. Similarly, customs at Mexican airports used to randomly inspect travelers before installing baggage X-ray machines in 2007 for more informed decisions.

Of course, in most situations there *are* ways to make better decisions than rolling the dice. The two most common alternatives to complex algorithms are human judgment and simple criteria.

- **Human judgment** is the precursor of algorithms. As discussed, empirically human judgment suffers from *additional* biases and therefore often is worse (i.e., more biased) than algorithms. On the flip side, humans have the ability to logically evaluate novel situations. For example, when the first Martian visitor arrives at Mexican customs, there will be no algorithm to decide whether his flying saucer should incur any customs duty but a human can come up with a credible argument that it is an electric device for personal use exempt from import taxes. There is also an interesting psychological phenomenon that many times humans seem to be more prepared to *accept* a human decision than a machine-made one. This idea was even baked into the European General Data Protection Regulation (GDPR) which prohibits "solely" automated decisions which have a "significant" or "legal" effect on an individual, unless they are explicitly authorized by consent, contract, or member state law. Note that the rules explicitly do *not* say "unless there is abundant evidence that human influence on the decision will introduce material harmful bias." Apparently even European lawmakers can exhibit overconfidence in human decision making!

- **Simple criteria** could be considered a very simplistic version of an algorithm; however, they are exceedingly transparent and therefore any bias they may entail will be explicit and hence can be openly debated. For example, tax authorities that only manually review a small fraction of tax filings sometimes have a practice to manually review and validate 100% of first-time filings of new companies. While this arguably is a bias against start-up entrepreneurs, it is a reasonable (and empirically validated) assumption that first-time filers are particularly prone of errors and therefore such differentiated treatment is warranted.

And you even can mix up things. **Hybrid approaches** combine an algorithm with human judgment and/or simple criteria. For example, an algorithm can flag potentially fraudulent tax filings but investigators can then prioritize the flagged filings and audit only a fraction of them. Likewise, banks use a credit scoring algorithm to divide applications into three groups: approved, rejected, and a "grey area" in the middle that is manually reviewed. A value criterion can often further refine the decision rule. For very small loans, a manual review may not be economical, so a bank may automatically reject applications for very small loans in the "grey" area. And for a tax payer with very large earnings arguably so much tax is at stake that tax investigators may briefly review *all* tax filings of very high earners to decide whether to audit; in this case, however, it would be useful for them to see the probability of tax fraud calculated by the score (psychologically it would anchor their judgment and therefore may help to fight some human biases, such as an ideological belief that all rich people are crooks).

The following alternatives to purely algorithmic decision-making give you three interesting design options for keeping biased algorithms at bay:

- **Avoid** algorithms altogether for decisions where their benefits do not justify their biases; this can occur because either an algorithm introduces biases that another approach can avoid, or its benefits are diminished because on an emotional level the algorithm's biases hurt more than whatever limitations the alternative decision process may have.

- **Restrict** algorithms to a subset of "safe" cases where either the risk or the impact of algorithmic bias is lower than for other cases.

- **Sequence** an algorithmic preselection with a human validation. This approach lends itself to asymmetric decision-problems where one outcome (e.g., a conviction of a crime) is considered severe while the alternative treatment (e.g., release of a presumed-to-be-innocent person from jail) is considered less severe; algorithms here might only have the power to select the benign treatment but will always be validated judgmentally before the severe treatment is selected.

We therefore need to understand how much better or worse an algorithm performs compared to other options. In the following section, we will discuss how to assess this.

How to Assess the Benefits of Algorithms

In order to assess if a biased algorithm performs worse or better than alternative decision approaches, the central question is which approach is more accurate. A simple measure of accuracy is **error rates**.

Error rates for binary decisions (e.g., predicting whether or not a loan will default or a prisoner will commit another crime once released from prison) can be measured with historical data. For example, I can collect data on all credit card applications a bank scored 12 months ago and then track for each customer whether he defaulted on any bank debt since. If I either collect data on actual decisions made by alternative approaches (e.g., for a while *The Wall Street Journal* gave monkeys darts to make alternative stock picks against which professional asset managers could be benchmarked—you can guess who won!) or simulate how alternative approaches would have decided historical cases, I can compare performance.

The **hit-ratio** combines two types of error rates: *rejecting* a customer who subsequently defaulted or *approving* a customer who subsequently repaid the loan are labeled correct decisions, and approving a customer who subsequently *defaulted* or rejecting a customer who subsequently demonstrated satisfactorily servicing her debt[1] are labeled false decisions (so-called false positive or false negative). There are two limitations with this metric, however: it counts both types of false decisions equally (whereas economically, one type of error might be a lot more costly than the other type—the loss from one defaulted loan usually is many times larger than the lost profit from rejecting one good loan), and if the algorithm produces a probability of an outcome, the hit ratio is also dependent on whether or not a prudent cutoff line has been drawn (e.g., up to which probability of default do you approve a loan? 1%? 5%? 10%? 20%?).

For situations where algorithms estimate the probability of an outcome, statisticians have developed much better metrics for measuring how well an algorithm rank-orders outcomes. A review by the Bank of International Settlement[2] (which develops standards for banks globally) recommends in particular the Gini coefficient. It doesn't live in a bottle—it's named after an Italian statistician, Corrado Gini—but like alcohol content, it is a number between 0 and 100 percent.[3] 0 means that a score is completely random (or the same for everyone) and hence as useless as vodka without alcohol, while 100 means that the score has perfect foresight. (A crystal ball maybe can achieve 100 but any algorithm predicting real-world phenomena with some degree of uncertainty has values below 100. For example, for a difficult decision problem such as assessing the probability of default of small businesses, you might find Gini coefficients in the range of 35-70.) A very similar metric that is very popular as well (possibly because it sounds like a brand of vodka) is the K-S statistic (which stands for Kolmogorov-Smirnov).

The best approach, however, is to simulate the actual economic outcome (i.e., profit or loss) of applying different algorithms or decision-making approaches.

---

[1]There are some intricacies to this case which are beyond this chapter but are important: a customer who is rejected by *all* financial institutions and therefore unable to take on any debt obviously can never *demonstrate* that he repays loans. One could argue that for lack of contradictory evidence, the algorithm's assessment that this person should not get a loan therefore was correct (in fact, the fact that all other banks seem to concur could be considered supporting evidence for this hypothesis). The result, however, could be a perpetuating bias against this type of customer, and we therefore will discuss in a later chapter how to generate data to bust such a bias.

[2]Basel Committee on Banking Supervision, "*Studies on the validation of internal rating systems. Working Paper No. 14*," 2005.

[3]Technically, Gini *can* be less than zero (but never more than 100); this happens if an algorithm is *worse* than random and you systematically can improve outcomes simply by doing the *opposite* of what the algorithm suggests. In this case, you simply turn your score upside down, and the Gini metric is positive again…

Take a bank: you may find an algorithm with a very high Gini coefficient (i.e., excellent predictive power) to be generally great but uniquely challenged in assessing very large loans (potentially due to a "big is beautiful" bias). The loss from approving a couple of very large bad loans very well might eliminate all the savings accomplished by the algorithm for smaller loans. To estimate the *economic* impact of an algorithm, you will need to obtain a historical sample of loan applications, a couple of business parameters (e.g., interest margins and late fees), and crunch some numbers, but the result from such a simulation—even if done in a crude, approximate fashion—can be eye-opening!

As noted earlier, algorithms are designed to be unbiased and therefore we should expect that generally algorithms outperform human judgment and more simplistic decision criteria if we benchmark them using the Gini coefficient or a more careful economic analysis. So when might we actually see algorithms perform worse?

- Algorithms require lots of homogenous data—a few hundreds of structurally comparable cases are the bare minimum, and some advanced machine learning techniques require hundreds of thousands of cases to unfold their power. If the amount of available data is small or the cases are not very homogenous, algorithms struggle due to overfitting—to the point that for very small samples or very heterogeneous data they perform worse than human judgment (which can use logical reasoning).

- If human decision-makers have access to information that is not available electronically (e.g., the body language of another person and other qualitative factors), they can outperform algorithms due to this information advantage. This will happen in particular if the approach to exercising such judgment has explicitly been designed to remove human bias or the decision maker has plenty of experience from having made hundreds of similar decisions with more or less immediate feedback.

If an algorithm performs worse than human judgment because of its bias, you also may want to consider how much of a **cost** and **speed** advantage the automated decisioning affords. If the cost of the algorithmic bias is small but the operational cost of removing it through human overlay is high, economically the biased algorithm still might be the best option.

Summary

In this chapter, we discussed the design of the overall decision architecture and to what extent there might be alternatives to using a (biased) algorithm for a decision. Key take-aways are:

- If an unbiased algorithm appears unachievable, the architect of the decision process should examine whether such an algorithm should be replaced by or combined with a non-algorithmic decision-making approach.

- Alternative options for the decision-problem at hand include treating all cases identically, making a necessary selection randomly, using human judgment, or using exceedingly simple (and hence transparent) criteria.

- Alternative decision making approaches—especially human judgment—often will be even worse (i.e., more biased) than an algorithm. The decision between alternative approaches hence should always be based on relative performance.

- The Gini measure is an excellent metric that can compare the quality of decisions of alternative approaches for binary decisions. An economic analysis of the cost of both false positives and false negatives is an even more meaningful analysis for comparing errors of different decisioning approaches.

- Empirically, algorithms perform best if there are lots (i.e., thousands) of homogenous cases; however, if data is rare (i.e., not even 100-200 historical cases) or a human decision maker can take into account additional information (usually qualitative in nature), algorithms may perform worse (i.e., suffer from more biases) than humans.

Knowing that all alternatives to an algorithm perform worse does not eliminate the problem that the algorithm may be biased, however, and you may still be exposed to legal, reputational, and business risks due to the bias. In the subsequent chapters, we therefore will focus on dealing with algorithmic biases properly.

Assessing the Risk of Algorithmic Bias

The last chapter laid the foundation of an informed decision on whether using an algorithm is a better or worse solution to a decision-problem than alternative approaches such as human judgment, a simple criterion, or rolling a dice. That discussion concluded with the observation that empirically, in many situations, algorithms make better decisions than alternative approaches: they make less errors (especially because human decisions often are even more biased) and can be both faster and cheaper. Introducing an algorithm to make decisions may come at the price of new biases, however.

In Chapter 15, we will discuss how to analytically establish whether an existing algorithm exhibits some form of bias. This can only be done once an algorithm has been developed, however. If you are still at the stage of designing an approach for making a particular decision and are merely considering an algorithm as one possible option, you want to understand the risk of algorithmic bias before actually investing in the development of an algorithm. This chapter therefore will lay out on a conceptual level how to discern situations with a high versus a low risk of algorithmic bias.

© Tobias Baer 2019
T. Baer, *Understand, Manage, and Prevent Algorithmic Bias*,
https://doi.org/10.1007/978-1-4842-4885-0_13

Insurance companies typically quantify risks by assessing the frequency and the severity of losses separately. A similar distinction is helpful here. We will discuss both what drives the likelihood of algorithmic bias and what drives the severity of a bias if it occurs.

Assessing Loss Severity

One approach to assessing the damage from biased algorithms is, of course, ethical considerations. For business decisions, however, risks often need to be quantified economically. The economic damage caused by algorithmic bias comes from three sources: legal risks, reputational risks, and model performance. Ethical considerations are reflected to the extent they are mirrored by regulations and public opinion.

- **Legal risks** refer to biases that translate into discrimination against individuals or cases that are banned by law. Regulations often draw hard red lines (e.g., the law may prohibit discriminating by gender or age), and violations of such laws and regulations can invoke substantial fines or even revocations of business licenses.

- **Reputational risks**, by contrast, are grey, fuzzy lines that can shift rapidly. Any sort of discrimination by an algorithm that is perfectly legal can nevertheless provoke the ire of the public if it is deemed unfair *and* evidence of the algorithm's bias gets spread, for example, through social media.

- **Model performance** is a pure **business risk**. An algorithm's bias can backfire, for example, if it triggers a vicious cycle or if populations shift towards classes where the algorithm has a harmful bias that causes business losses.

Quantification of these risks will often be colored by an institution's risk appetite. Aggressive risk-takers may choose to brush up against the limits of the law and take calculated gambles on reputational risk (e.g., by betting that a certain bias won't become known to the public); cautious organizations often play it safe by staying very well within legal limits (often interpreting the law even more strictly than courts) and minimizing any reputational risk. And model performance risk may be partially mitigated by frequent and rapid model performance measurements that allow an institution to quickly remedy a harmful bias if it shows up in decision results.

Assessment of the risks therefore will involve your knowledge and assessment of legal, reputational, and business implications of algorithmic biases in your concrete business context. An example for a very low risk context would be a typical marketing model deciding which customer to send a marketing offer. Assume you are an online bookseller and you use an algorithm to determine which customers get a marketing email to pitch this book. Obviously *everybody* would want to read about this exciting topic—but if your marketing algorithm discriminates against people who have bought multiple knitting books in the past and incorrectly assumes that sock-knitting grandmas are not deeply concerned about algorithmic bias too, it's unlikely to put your livelihood as a bookseller at risk. Whenever you optimize between "innocent" choices, algorithms, especially the low-cost ones generated by machine learning, promise substantial upside with close to no downside.

On the other side, some uses of algorithms can have extreme downsides. One very controversial area is criminal prosecution. Decisions such as whether to set a suspect free on bail or whether to parole a prisoner are classic decision-problems that would lend themselves to statistical algorithms, and there is ample evidence that if such decisions are made by humans, bias can creep in. If, however, such an algorithm had any bias of its own, it would obviously constitute a grave violation of fairness and democratic principles, and a healthy debate can be led on to what extent such decisions should be made by an algorithm at all, given that algorithms by design do not consider "unique" circumstances of a given case. Today most societies choose a manual (judicial) decision process, although one could argue that a hybrid approach where a judge's decision is grounded in a non-binding recommendation by a carefully calibrated algorithm might be even better if it helps to counteract any biases found in judges' decisions.

Other areas where biases would come at high cost are medical decisions and credit underwriting. For example, in the US, credit underwriting is tightly regulated, and discriminating against protected classes of customers not only violates democratic ideals but carries heavy penalties. If there is even the slightest risk of the algorithm exhibiting a bias against a protected class, the use of an algorithm may be conditional on the data scientist being able to prove that no illegal discrimination takes place. This may have important implications for the data scientist's choice of modeling techniques and data.

And for a hospital, it could be an extreme reputational risk if a patient dies because an algorithm denied a critical medical procedure even though most practitioners would consider the patient a typical case warranting said procedure. Even if an algorithm suggests a particular course of action with an extremely high degree of confidence (e.g., suggesting that a particular procedure has a 99.9% probability of failing), a human decision-maker may decide to go ahead with the procedure if it is considered the standard treatment in this situation and denying it would impose unacceptable reputational (and potentially also legal) risks.

The size of the business risk is usually driven by the amounts at stake, which is why most institutions limit automatic decisioning by algorithms by transaction amount (e.g., amount lent, invested, or invoiced). More sophisticated institutions adjust the amount for objective moderators of risk. For example, a bank might adjust the loan amount for collateral value, while for financial investments, more complex risk metrics such as value-at-risk can be used. In trade finance, simple criteria are used to demark algorithmic decisioning, such as whether a high-risk country is involved or if the two trade parties have done transactions before.

Assessing the Propensity of Bias

The question of how likely an algorithm is to exhibit a harmful bias is independent of the assessment of how severe the consequences of a bias would be. Whether you assess loss propensity (i.e., the likelihood of a harmful bias to occur) or loss severity first can be decided based on convenience; often, however, loss severity in the end decides the course of action, and it therefore may be most efficient to assess severity first.

A comprehensive assessment of an algorithm's propensity to be biased would naturally have to review the full list of algorithmic biases discussed in the second chapter. For an expedited review, I suggest the following pragmatic three-item check-list:

- Is the algorithm replacing a real-world decision process that is known to have suffered from biases in the past (e.g., gender or racial bias in hiring decisions)?

- Is the historical data used for developing the algorithm extremely sparse (e.g., below 200 cases or less than one-tenth of the sample size data scientists typically use)? While there are relatively robust modeling techniques, such as logistic regression for a sample of, say, 600 observations, many machine learning techniques such as tree-based approaches literally would blow up with such a small sample.

- Is the historical data used for developing the algorithm materially compromised? For example, the sample selection has been biased (e.g., cases have been selected by humans suspected to be biased), or specific subsegments of the population that are expected to behave differently from the rest are missing, or the period from which the data has been taken has been materially different from the situation expected for the future.

If you answer any of these questions with yes, there is a high risk of bias. What does this imply? Unless the severity is negligible, you may first want to pursue specific measures to counter the bias, such as running an experiment in order to be able to collect unbiased data (which is the subject of the subsequent chapters), and introduce an algorithm only if you are confident that any harmful biases have been removed. If this can't be accomplished, you should consider whether safer options for the decision-making than an entirely algorithm-driven process exist.

Summary

In this chapter, you explored how to quantify the risk of a specific algorithm being biased in a forward-looking manner. Key take-aways are:

- You should assess both the severity of losses you were to face if a bias materializes and the likelihood that the algorithm actually is biased.

- Loss severity is driven by legal, reputational, and business risks.

- Assessment of loss severity reflects both your knowledge of the specific business context and your risk appetite (which acts like a discount factor applied to worst-case scenarios).

- The top three indicators for a high propensity of bias are 1) human biases present in historical outcomes that inform the learning of the algorithm, 2) sparse data as a source of statistical instability, and 3) material limitations of the historical data sample that render it biased.

- If loss severity is negligible, algorithmic bias may not pose a serious concern, and a biased algorithm may still be employed, with future updates of the algorithm possibly attempting steps to reduce or remove the bias.

- If loss severity is medium, an economic cost-benefit trade-off may find it preferable to first take steps to remove the algorithmic bias before deploying the algorithm.

- If loss severity is high, not only must every possible step be taken to avoid algorithmic bias but even if the residual rest of algorithmic bias is very small, the architecture of the decision process may impose additional safety mechanisms such as human validation of algorithmic decisions.

You now are in a position to take a much more nuanced view of algorithmic bias: rather than seeing algorithms in a black-and-white world where things are all good or all bad, you know that algorithms have benefits, risks, and costs, and that sometimes even a biased algorithm is better than no algorithm at all.

You also know that sometimes algorithmic bias cannot be eliminated, so let's now discuss how to deal with biased algorithms in order to minimize the damage they cause.

How to Use Algorithms Safely

In the last chapter, you learned how to assess the risk of a particular algorithm being biased. The conclusion was that in many situations, we may find that a certain risk of algorithmic bias is present but that based on a cost-benefit analysis, the algorithm will still make better decisions than other approaches (such as even more biased humans). This situation can be compared to a life-saving medicine with serious side-effects. Just as the doctor will try to find ways to alleviate the side-effects of a medication, in this chapter we will discuss what steps you can take to protect yourself from algorithmic bias.

How should laymen deal with algorithms in order to avoid problems from algorithmic biases? The most important line of defense for laymen is to use algorithms in an informed fashion.

By analogy, consider how an informed consumer thinks about buying groceries and consuming food: few of us are professional nutritionists or have the necessary medical and biological knowledge to understand the health risks of specific food items. Even worse, if you follow the news flow on the "latest" insights on what is good or bad for you, you sometimes cannot resist the impression that even science is still figuring out what specific food items do to

© Tobias Baer 2019
T. Baer, *Understand, Manage, and Prevent Algorithmic Bias*,
https://doi.org/10.1007/978-1-4842-4885-0_14

our bodies. However, there are broad and basic rules of thumbs that help us to nevertheless make informed decisions. For example, we know that fiber is good but that excessive sugar is bad, and we are aware that balanced nutrition is a good hedge against major nutritional missteps.

What a consumer should *not* do is to blindly buy whatever tastes good—we know that we are almost guaranteed to end up with too much junk food—and we also know that we should not be too naïve about advertising. A critical habit of informed consumers therefore is reading—of product labels and also of the occasional article on the subject.

By reading this book, you are obviously already on a golden path towards becoming an informed user of algorithms. Specifically, I would like to urge you to adopt three habits:

First, ask, ask, ask. Pepper your data scientists with curious questions. *Curious* questions are questions that purely help your understanding. Don't assume that your data scientist is evil or hiding facts from you—that only will poison the relationship and tempt your data scientist to become defensive—but to the contrary, let your natural curiosity drive the conversation. Ask how exactly to read the output of the algorithm, and honestly marvel at the fact that the new credit score can pinpoint a group of ultra-safe companies that only have a 0.03% probability of not repaying their debt. Ask how the credit score can achieve this without reading your carefully crafted credit memos, and pick your data scientist's brain about what might happen if a company forges its financials. Most importantly, ask matter-of-factly what biases the algorithms might have, in which situations you should be particularly careful with the outputs, and what it may take to derail the algorithm. What you want to achieve through this is to help your data scientist to better understand the challenges the algorithm might encounter in the real world and what might be the greatest weaknesses of the algorithm. This allows the data scientist not only to think of improvements for the next version of the score but also to better monitor the algorithm and to protect business outcomes by suggesting specific constraints on the use of the score. The more your data scientist sees you as a partner rather than a critic or enemy, the more productive this relationship will be. And don't pretend to be a data scientist yourself or be scared off by not being one—your value add is exactly that you approach problems from a totally different, non-technical perspective that is therefore complementary to the data scientist's statistical perspective.[1]

---

[1]An old joke might illustrate this: after a door-to-door sales rep extolled the many technical strengths of his product, the latest and most advanced vacuum cleaner on the market, he emptied a bag full of dust on the floor and enthusiastically exclaimed that he would eat any dust from the floor his vacuum cleaner could not vacuum away. His customer, a farmer, handed the rep a spoon and asked, "And how will the vacuum cleaner work given that we have no electricity in the house?"

Second, understand for which cases the algorithm has an insufficient basis for a useful prediction, and ask the data scientist to assign a "don't know" label to them instead of an estimate. It is actually a potentially fatal design issue that many algorithms will *always* provide you their best estimate (even if it is just the population mean or a random number) rather than honestly stating "I don't know."[2] Labeling certain classes of cases as "don't know" is a double smart move. On the one hand, it enables superior hybrid decision processes—for example, these cases lend themselves to some human intervention or a conservative decision rule. On the other hand, it also prevents human users to be biased (anchored) by an unreliable estimate—as soon as an algorithm provides an estimate, there is a risk that this number develops a life of its own and will subconsciously bias any overriding human judgment even if it is perfectly clear that this number is purely random. Often that effect is amplified by formatting—an output of 2.47% looks *very* precise even if it is simply the population mean. Algorithms don't even round their wildest guesses! The only output that can appropriately caution human users is "I don't know."

Third, ask to regularly see meaningful monitoring reports, and if you believe that additional metrics would help you in assessing whether everything is working, try to add them to the report. Just as thoughtful nutrition should be complemented by regular health checks, and high cholesterol results trigger a review of your eating habits, the factual nature of monitoring is invaluable in detecting algorithmic bias (as well as many other problems of algorithms). In fact, monitoring is so important that in the next chapter, we will review in detail how to monitor algorithms. Just remember that these monitoring reports must be *meaningful*. This means, first of all, that all metrics actually need to mean something to you (otherwise they just waste your time); secondly, that taken together they should be rather comprehensive; and finally that the report should draw your attention to the important insights—it doesn't help if the one metric screaming "bias!" is buried amongst a thousand other metrics that all look good. Good reports therefore are designed to draw your attention in particular to two types of situations: metrics that are outside of the value range considered "safe" or "OK," and metrics that show big changes. This is actually also how our brain constantly monitors for dangers—it looks for unusual things (you already encountered the bizarreness effect) and it looks out for big changes. The experience of one of my banking clients illustrates the power of monitoring: one very meaningful metric for home equity loan performance, the scaling factor of so-called vintage curves, showed big changes already in 2005—a full two years before the global financial crisis broke out. This metric was the equivalent of an early warning signal for

---

[2]Note that simple algorithms often will throw up an error if even one input field is missing. At times this is actually useful; however, because in many such situations it is still possible to produce a useful estimate, more advanced algorithms perform so-called missing value imputation—but this errs on the opposite side and you will even get an estimate if *all* inputs are missing!

tsunamis (and will be revisited in the next chapter when we discuss calibration analysis).

Taken together, these three practices not only render users of algorithms informed about risks from algorithmic bias without requiring any particular technical knowledge but also drive to specific solutions.

Understanding the specific risks and limitations of algorithmic biases through conversations with data scientists (e.g., data weaknesses) can inform precautions taken on the business side (such as manual review of certain types of cases and other limitations on the use of the algorithm) as well as the modeling side (such as specific corrections to the data, removal of specific variables from the equation, or targeted monitoring of particular metrics). And effective monitoring helps users detect new biases arising or existing biases growing.

One final note: Among the three habits recommended, the second one (getting the algorithm to admit if it really does not know) is the least common practice. This is rather ironic given that Socrates remarked that he knew more than his fellow Athenians because he *knew* what he did not know ("Ἐν οἶδα ὅτι οὐδὲν οἶδα," "I know that I know nothing")—as opposed to others not even knowing what they don't know. Hopefully this book will change this!

Summary

In this chapter, you learned three habits laymen users of algorithms can adopt to get a handle on risks of algorithmic biases without having to be a technical expert. The key points to remember are:

- Discussing and understanding the risks and limitations of algorithms not only enables laymen users to design appropriate safeguards in business processes but also to become a thought partner to data scientists with highly complementary real-world business insights.

- It is vital for business users to neither make data scientists defensive about their work nor to allow technical jargon to get in the way of discussing real-world risks in pragmatic business terms. An amicable discussion between business users and data scientists is immensely valuable for both sides.

- One very practical outcome of a discussion of an algorithm's limitations and dangers of bias is the identification of objective criteria for cases where algorithmic outputs should be suppressed and quite literally replaced with "I don't know" labels.

- Suppressing numeric outputs from appearing *anywhere* where users should read "I don't know" is critical to prevent anchoring of users in a biased value.

- The other very practical outcome of a discussion of an algorithm's limitations and dangers of bias is the definition of a tailored monitoring regime.

In the next chapter, you will dive deeper into the topic of monitoring, considering not only proactive monitoring as a way to prevent known weaknesses of an algorithm from blowing up but also as a technique to detect unexpected biases in algorithms that conceptually appear to be sound. Subsequently, you will explore more broadly managerial strategies for dealing with biased algorithms.

How to Detect Algorithmic Biases

In the previous chapter, I indicated that monitoring plays a central role in managing algorithms. This is surprisingly tricky. As Ron DeLegge II put it nicely, "99 percent of all statistics only tell 49 percent of the story."[1] As a result, a lot of rubbish is said and done because of meaningless numbers showing up in some report. Even if no bad intentions are involved, a poorly calculated or interpreted number can seriously mislead you. This chapter is a comprehensive review of how best to monitor algorithms for biases from a user's perspective.

I have tried to make this chapter interesting—we'll talk about gorilla men, drunk cooks, eating guinea pigs, beer, and Martian ghettos, among other things—but it's still about monitoring algorithms. So if you believe you will never monitor an algorithm in your life, feel free to skip this chapter. Then again, you have no idea what unexpected joys you will miss—so maybe you want to read on anyhow! I think about monitoring an algorithm similar to a medical check-up. Doctors regularly scan their patients for standard metrics that can indicate hidden health issues—think cholesterol levels in the blood

---

[1]Ron DeLegge II, *Gents with No Cents, 2nd edition*, Half Full Publishing Group, 2011.

© Tobias Baer 2019
T. Baer, *Understand, Manage, and Prevent Algorithmic Bias*,
https://doi.org/10.1007/978-1-4842-4885-0_15

and spots on the skin that may be cancerous. However, doctors also perform customized checks for known health issues (e.g., to ensure that the drug dosage given to an epilepsy patient continues to suffice) and to follow up on new warning signals that come up in regular health checks (e.g., a biopsy of a suspicious spot on the skin).

In doing this, we will have to deal with two particular challenges. First of all, algorithms can sometimes mirror the biases occurring in the real world, as discussed in Chapter 6. We therefore need to find a way to distinguish real-world biases mirrored by an algorithm from biases introduced or amplified by an algorithm because the implications and remedies are very different.

Second, the introduction of machine learning has complicated monitoring quite a bit. This is due to two reasons: machine learning models are a lot more complex and less transparent than simpler, traditional algorithms such as a logistic regression, and machine learning models can be updated a lot faster than man-made models, sometimes as frequently as once a day (or even continuously, in the case of real-time machine learning). This challenges algorithmic monitoring regimes. And not untypical of technological progress, the state of the art in the development of algorithms has progressed much faster and further than the state of the art in monitoring them.

In the following sections, we therefore will first review the basics of monitoring algorithms. We'll then briefly discuss how to understand the root cause of a suspected bias that came up in our monitoring, including the comparison of algorithmic biases with real-world biases. Finally, I'll propose specific monitoring approaches for "black box" machine learning models and frequently updated algorithms.

Monitoring Algorithms: The Basics

The purpose of monitoring is to be alerted of likely problems—similar to how sudden pain in the sole alerts us of a likely injury, such as having stepped onto a sharp object. This is actually a three-step process—first, we need to define which metrics we want to track and (importantly) which value range we consider "normal" (versus an indication of a possible problem) for each metric (otherwise we just produce meaningless numbers); then we need to build the process and routine to periodically calculate the metrics we have chosen; and finally, we need to go through each report, identify which metrics are actually outside of their normal range, and then assess and decide what to do. Many modeling reports miss half of this—they produce pages and pages with numbers whose meaning eludes most readers, and there is no follow-up action; as a result, warning signals can hide in plain sight.

As we define a couple of meaningful metrics, it is helpful to differentiate two types of metrics: forward-looking versus backward-looking metrics. This issue arises because algorithms used for decisions usually are predictive—they provide an estimate for something that we don't know yet. Many times (but not always) the truth will reveal itself in the future. For example, if an algorithm approves a loan with a one-year term, one year later I will know for sure whether the customer has repaid the loan on time, unless the customer decided not to take the loan (e.g., because my algorithm suggested such a low probability of repayment that my risk-based pricing model calculated such a high interest rate that the customer immediately left my branch, swearing all the way to the door), in which case I never will know whether the customer would have repaid this loan—a problem we'll discuss in due course.

Forward-looking metrics can be calculated when I use the algorithm for making a business decision. For example, I can calculate the approval rate for all loans applied for this month and compare it with my target range. Backward-looking metrics are the most direct indicator of a problem but they can only be calculated with a delay (once the truth has been revealed—e.g., a year from now I can calculate the default rate of all loans given this month and compare it with the predictions made at time of origination).

A major issue with backward-looking metrics is that my decision-process often will bias my data. For example, a year from now I only have performance data on loans I approved but not performance data on loans I rejected. Similarly, I can measure the sales performance of sales people I hired but can't generate data on the sales performance of applicants I rejected. In order to truly analyze the performance of an algorithm and thus detect a bias in rejecting certain classes of cases, I need to think of ways to find data on such cases—so-called *reject inference*. Sometimes useful data can be found externally (e.g., in some cases, I may be able to obtain from a credit bureau information on whether a rejected customer has repaid a loan that she obtained from a competing bank); sometimes I need to generate such data myself by randomly selecting a sample of "rejected" applications and providing a loan nevertheless just to see what happens. We will revisit the topic of generating unbiased data in a dedicated chapter soon.

In the following, I will present a mix of simple and more advanced metrics. All metrics—but especially simple ones—come with limitations; however, as long as you keep the doctor analogy in mind and realize that a warning sign is just that (a sign that something *might* be wrong but not a guarantee that there is a problem with 100% certainty), I do believe that simple metrics have a lot of value in spite of their limitations.

My two favorite forward-looking metrics are the distribution analysis and the override analysis.

- *Distribution analysis* looks at the distribution of algorithmic outputs by certain case attributes that I hypothesize to be relevant for a bias. For example, I can segment applicants by gender, age, or city and calculate respective approval rates. If I approve 40% of men but only 3% of women, I may experience a serious bias. And I reiterate: it merely means that I *may* experience a bias. There also could be perfectly acceptable reasons. For example, imagine that you are running a budget airline that charges for flights by passenger weight—not unreasonable given that heavier passengers will burn more fuel, possibly eat more, and if you have done away with seats and just allocate a couple of inches on benches, heavier passengers also may get more bench space. If you charge on average more for men than for women, do you discriminate against men? Not if they are on average heavier than women—if you look at a sample of *just* passengers weighing exactly 165 pounds, you may find that you charge men and women *exactly* the same fare. The distribution analysis here therefore is only a first step—it is a quick "dip test" to check whether there *might* be a bias. In a second step, a more careful analysis can and should be conducted if you have reason to believe that harmful bias might be present.

- *Override analysis* is applicable where my decision process entails some element of override—for example, because a human validates algorithmic decisions or because individuals affected by a decision can lodge an appeal. On an absolute level, a high override rate is a first indication that an algorithm might have a problem; a deep-dive analysis (override analysis by segment or a root cause analysis) can indicate whether overrides are concentrated in a particular group of cases. And if the override process captures the override reason in a meaningful way, I may even be able to pinpoint which particular model input or logic causes a bias (e.g., if I screen CVs automatically but HR analysts manually review cases with a borderline score, they might record that they mostly override rejections where the algorithm seems to have penalized graduates from a non-Ivy League school—this suggests that my algorithm might suffer from an unwarranted preference for Ivy League schools).

My two favorite backward-looking metrics are the assessment of calibration and rank ordering.

- *Calibration* is the litmus test for whether an algorithm estimates correctly what it is supposed to estimate. It compares predictions with actual outcomes (hence it is backward looking). Recall that algorithms aim to be right *on average*, especially where outcomes are binary events and hence the probability for an event happening even conceptually cannot be validated with a singular case. So for a **binary event**, we compare the average probability given by the algorithm (e.g., for a portfolio of 1,000 loans, the algorithm may indicate an average probability of default of 2.3%, which would imply 23 default cases) with the actual percentage of cases where the event in question has occurred (e.g., to my dismay I may have suffered 472 defaulted loans. which means that something went seriously wrong!). For a **continuous outcome**, I similarly compare predictions and outcomes. For example, if my algorithm predicted on average 107,233 hairs per person for a famous Italian soccer team and a careful count yielded in average 107,234 hairs, then my algorithm can be said to be spot on!

- *Rank ordering* is very different from calibration. Recall that the purpose of an algorithm is to give some fair, fact-based guidance on how to *differentiate* treatment between different people. If you are in the business of buying people's hair to produce wigs and your profits depend on how many strands of hair you get, then an algorithm that indicates an estimate of 107,233 hairs for *every* person is rather useless—you need to know who has little and who has lots of hair. In Chapter 12, you encountered a metric to measure rank ordering for binary outcomes: Gini.

Let's now consider in a bit more detail how to perform each of these analyses (this is where the devil hides) before discussing how to decide on appropriate "normal" ranges for each metric—which is how we can decide when to conduct a root cause analysis to follow up on anomalies. In doing so, we will also briefly touch upon a couple of alternative metrics that overcome specific limitations of the basic analyses discussed above.

How to Do a Proper Distribution Analysis

This part of the book is aimed at users of algorithms, most of whom will not be data scientists—nor might ever had any inclination to get anywhere close to a statistics book. Therefore I will not go deeply into statistics here but suggest only analyses that either can be done with basic means such as a spreading software (e.g., MS Excel) or that a user could request a data scientist to produce.

Also, I want to clarify that I use "distribution analyses" somewhat loosely:

- For multinomial outputs (i.e., the algorithm suggests one of several possible *categorical* or non-numeric values, such as "which book to recommend next"), distribution analyses would measure the relative frequency of each category/possible output value—and if there are too many of them to fit into a neat little table, the report may just list, say, the top five and summarize the others by category (e.g., fiction versus non-fiction books).

- Continuous outputs can be summarized in a single metric (e.g., the average or median), and this often is the easiest for users to digest (e.g., if you are a bookseller, you probably would be alarmed if you saw that the average price of the books recommended by your algorithm trends downward—this bias to thriftiness might put you out of business!); however, sometimes it is more meaningful to define value ranges (e.g., five broad ranges of book prices) and track how the (percentage) distribution of cases across the five ranges shifts over time (e.g., if you find that your algorithm is recommending less and less medium-priced books and instead is peddling either extremely cheap or extremely expensive books, the average price per book might stay constant but you still have something very peculiar going on that may or may not be in your best commercial interest). When reporting ranges of continuous values, there is also tremendous value in explicitly showing the minimum and maximum value, respectively, that occurred in the data because so-called outliers can be both a cause and a symptom of biases.

- For binary outputs (yes/no outcomes) your natural choice is the calculation of a percentage (i.e., a single metric), although you may want to keep an eye out for situations where really a third category ("inconclusive") lurks behind the simplistic black-and-white view of the world and hence the distinction of three categories would be a lot more meaningful. For example, if one year after loan originations many customers lurk in the twilight of being 60-89 days behind, they technically have not defaulted (yet) but I would be very hesitant to claim that they are "good" customers.

Even if we keep things simple, however, it is important to know what to ask for and how to interpret results. I therefore want to introduce four important concepts:

- The distinction of flow and stock numbers

- Significance testing

- Marginal significance testing where you control for other explanations (unrelated to biases) of differences in outcomes

- Materiality

First of all, it is critical to be thoughtful if you are looking at stock numbers or flow numbers. Imagine a loan portfolio. Flow numbers relate to new loans that are opened (i.e., they are flowing into the portfolio); stock numbers relate to the total stock of loans in your portfolio that have originated at various points in time. If you deal with mortgages or other loans that are active for many years, some loans may be very old; other loans literally might have been booked just yesterday.

I briefly mentioned that stock numbers can be tricky because they can introduce statistical artifacts—for loans originated just yesterday, it is technically impossible to already be 90 days past due (which is the most typical criterion for labeling a loan "in default"); by contrast, a bunch of loans that have been on your books for five years already had plenty of opportunity to default, hence the percentage of loans that are in default will be much higher. Comparing those old loans with new loans does not make any sense at all—we need to compare apples with apples. For this reason, I always recommend to analyze flow numbers.

The second concept I want you to keep in mind is the concept of statistical significance. If you come home tonight to find that your five-year-old has cleaned up his room, does this mean that he suddenly has subscribed to the concept of perfect cleanliness, so messy rooms are a thing of the past? You'd wish so—but you probably know better. You only would start to believe that your son is a miracle if you see a consistent streak of a clean room evening after evening. The same applies to data—data may not willfully try to put you in a good mood to get some ice cream or approval for a sleepover, but it still may look good or bad out of a fluke. Statistical significance is a very sophisticated answer to the question, "How many times do I need to see something happen in order to believe that the world has changed for real?"

Your data scientist should be your go-to person to put numbers to this. If you merely want to compare two averages, the **t-test** can tell you how likely it is that a difference about which you agonize is a fluke.[2] For example, you notice that this week your CV screening algorithm has picked only one woman out of seven applicants (a paltry 14%), while last week you had three out of five (an enlightened 60%). Does this mean that your algorithm has developed a harmful bias? It's a tough call because the ratio of women fell by 75%, but at the same time, we're talking about two women less and two guys more this week compared to last week, which very well might be fluke.

Your data scientist friend will be able to run a t-test in your standard Excel within less than a minute. The result might indicate a P-value of 20%, which means that if nothing changed and your algorithm continues to favor women with the same probability as before (this is called the null hypothesis), you still have a 20% chance of the numbers coming out exactly as you saw them. Statisticians typically look for a P-value of at most 5-10% to give any credence to outcomes, and in order for a result to be "highly significant," you would want to see a P-value of not more than 0.1-1%. In other words, the handful of CVs you have seen is rather meaningless from a statistical perspective.

This, in turn, is helpful to consider when deciding how much data you need to accumulate before you can run a meaningful analysis (e.g., if you should "save up" all applications over a period of a day, a week, or maybe a quarter before you look at approval rates). As a very, very rough rule of thumb, below 30 observations of anything, it is very hard to distinguish a fluke from real trends. (Even though in social sciences this is done all the time—which in turn led to the so-called replication crisis in the field of psychology—so trust me on this![3]) Larger samples are always better, and in general I aim to have at least 100 cases, so I rather would choose quarterly intervals for my analysis if I then can have 100-200 cases than monthly intervals with 30-50 cases.

Third, I want to share with you a bit of statistical wizardry your data scientist friend could help you with that sheds a light on algorithmic biases in a much more effective way than your simple population averages.

---

[2]As this is neither a book about statistical hypothesis testing nor a section of the book squarely aimed at statistics pros, I purposely don't discuss details such as whether to use a one- or two-sided t-test or if a z-test would be better. I refer users to the trusted hands of their data scientist for choosing the best variation of these tests in light of the specific circumstances. The good news is that if you want a directional marker of a problem, all of these tests will wave at you if something is really fishy—just as when you smell rotten eggs, it doesn't quite matter if you sniff through your left or right nostril and if you hold the egg 1 inch or 10 inches away!

[3]Or look up J. Cohen, A Power Primer. *Quantitative Methods in Psychology*, 112(1), 155–159, 1992, for authoritative guidance on sample sizes.

Let's revisit the airline pricing example. You had an interesting start-up airline that charges tickets by body weight—but of course, ticket prices also reflect the distance flown, whether the ticket was bought the day before departure or three weeks in advance, and how many seats were still available on the plane in question at the time of booking. You also had a real concern that your airline might discriminate against men (not least because its entire management board consists of women!).

If you are keen to prove that your airline doesn't discriminate against men, you can share your algorithm, and anyone can verify that the passenger's gender does not go into the price calculated by the algorithm. However, the airline's sales staff (all women, too) has the power to give discounts when selling tickets over the phone or in the airport. Your data scientist friend therefore can run a regression test that tries to explain the final price paid by the passenger by just two variables—the output of the algorithm and the passenger's gender. If this analysis shows that both factors are *significant* in predicting the final price paid, you have pretty strong proof that your airline *does* discriminate against men. You also have found that in this case, the problem was not the algorithm itself but the human interaction that the airline allowed to modify the algorithmic output (maybe male passengers flirted while only female passengers drove hard for a bargain?). This is an important insight—very often biases are *not* caused by an algorithmic bias but by a judgmental overlay allowed by a hybrid decision approach. Given what we know about the biases of human behaviors, this is not surprising at all.

Now imagine, however, that the pricing department of your airline is a lot more cagey and less cooperative. The public relations office insists that apart from the objective factors mentioned above, no passenger attributes are taken into account—but refuses to share the algorithm with anyone. However, it turns out that a major travel agency has data on specific tickets it bought from this airline that can be used for some statistical analysis. In this case, your data scientist friend can do the following: he first will build a pricing model to estimate the ticket price using all available factors except gender of the passenger. Essentially, he tries to reverse-engineer the pricing algorithm of the airline. Let's call this first model the "objective" price of the ticket. In a second step, he runs the two-factor regression where he predicts the price paid by the passenger using the objective price and the gender of the passenger. Again, you want to know whether gender turns out to be a *significant* factor in explaining price. Since here the tickets were bought from a machine (the travel agent will query the computer of the airline with no human interaction on the airline's side involved), you now have proof that the algorithm of the airline *in effect* discriminates against men. Note that I had to qualify this last statement with "in effect;" the reason is that you don't know if gender is a direct input in the pricing algorithm or if the effect is an indirect one.

Fourth and last, I don't want to mention the concept of significance without also mentioning *materiality*. Significance is the first line of defense—*insignificant* phenomena have a high risk of being just fluke and noise, and therefore should not inform decisions. On the flip side, however, people sometimes get so excited about a "significant" result that they don't notice its immateriality. Especially if you have very large data sets, very small absolute changes could become statistically significant. For example, with enough data, even the average ticket for men just being 23 cents more expensive than tickets for women or a 77.1% acceptance rate for female candidates compared to 77.3% for men might be statistically significant. In a situation like this, the question is whether trying to fix this is really the best and most impactful use of your time—especially if you consider that the absence of more blatant differences suggests that there is no strong flaw in the system and the relatively small differences may be unwanted effects that could be very difficult to fix. I don't imply that this is a matter of black and white (if you are running a country with a billion people, 77.1% versus 77.3% might affect a million women)—my only point is that before taking action because of a *significant* finding of bias, have a look at the *materiality* of the bias as well and make an informed decision that the size of the effect is worth your time and efforts!

How to Do a Proper Override Analysis

When looking at overrides, you are trying to assess three aspects:

- Is the absolute level of overrides too high?

- Are there "hot spots" of overrides that might signal specific issues with my algorithm?

- Do the reason codes point to specific issues of my algorithm?

There is no absolute benchmark for what override rate is good or bad. However, if you consider that algorithms are designed to eliminate a lot of harmful biases, to save cost, and to speed up the process, it is fair to expect that algorithms get the majority of decisions right. Therefore, as a general rule of thumb, I don't want to see an override rate of more than 20%, and in many cases (e.g., high-performing credit scores—refer here to the measures of rank-ordering capability discussed in the next section for how to define "high-performing") I often have seen override rates way below 5%. In general, an immediate warning sign is a significant rise in override rates, and a secondary warning sign is a stable override rate that appears to be much higher than what you would expect based on comparable situations.

The identification of "hot spots" can go two ways. The brute-force approach is to simply look at override rates by individual segments (e.g., men versus women) and identify any segment with a significantly higher override rate. This is simple and practical if you are interested in a bias against a limited number of well-known segments. On the other hand, if you are more agnostic, you can ask your data scientist friend to run a little decision tree to identify the segments with the highest override rate (this is a statistical model where the yes/no flag of whether an override has occurred is predicted using all the other variables you have available). The tree will visualize exactly where the overrides are concentrated—for example, you might find "if the applicant is Martian, over 150 years old, and owns more than three saucer-shaped vehicles, over 90% of applications see a manual override."

Such hot spots are not a proof of a bias per se, but they tell you exactly where to start your investigation. For example, you might find that your algorithm is rejecting all applicants of the above-mentioned class of Martians—maybe due to one of the issues discussed in Part II that can arise if the modeling data is biased or a particularly class of applicants is very rare in the data. In this case, you would be able to point out to your data scientist exactly what issue to fix in the algorithm. Of course, it also might be a case of biased judgment by your case workers—in which case the algorithm might be perfect and the solution would be to better train the humans who make override decisions.

The analysis of reason codes depends very much on the structure in place for collecting meaningful reason codes. In loan applications, I frequently observed credit officers overriding a rejection by the loan scoring algorithm because of "insignificant delinquencies" reported in the credit bureau. Here the algorithm had developed a neurotic personality: if an applicant already owes a large amount of money to another bank and is late in servicing this debt, this obviously is a serious warning sign. However, many times one can see in the credit bureau that the customer is servicing all debt on time except for a tiny item, often less than 1 US$. The story behind these tiny items usually has nothing to do with credit risk—sometimes it is a system error caused by rounding, and sometimes it is a disputed minimal fee that the customer refuses to pay out of principle (e.g., she closed the account but the bank charged another monthly account fee). Many credit scoring models capture such delinquencies through a very crude method called a *dummy*—a binary variable that is "yes" if there is *any* delinquency and "no" otherwise. When I discovered this problem through this override analysis, I advised the bank to replace the dummy by a continuous variable that measured the delinquent *amount relative to a reasonable benchmark* (e.g., the total debt of the customer). Such a variable removes the algorithm's neurotic bias against tiny delinquencies while actually being more effective in capturing the warning articulated by large delinquencies.[4]

---

[4]This is because the neurotic aspect of the variable—ringing an alarm bell when the customer actually was perfectly good—had diminished its predictive power.

How to Assess Rank Ordering of a Score

Before we dive into calibration, let's discuss rank ordering because I believe that this will aid the understanding of the more advanced analyses of calibration for laymen.

Earlier, we already got to know the Gini coefficient to assess the rank-ordering ability of an algorithm for binary outcomes. A very similar but not entirely identical measure that is used a lot is the Kolmogorov-Smirnov statistic (or simply the K-S statistic). It looks and feels like Gini (and also comes on a scale from 0 to 100) but for the same algorithm, K-S scores tend to be a bit lower, at times up to 10-15 points (e.g., where the Gini is 50, K-S may be in the 37-42 point range). Both statistics are calculated on a sample where you know both the prediction of your algorithm and the actual outcome. I personally prefer Gini—whereas K-S basically measures a single point in the distribution of predictions and outcomes (namely the point where the algorithm is "at its best"), Gini incorporates every single prediction and so is less forgiving if, say, for a certain range of predictions (such as the estimate of hair for very short people) the algorithm is weak, which also explains why the numbers are somewhat different. However, both metrics are similarly useful—the most important recommendation is to use only one of them because then you always compare apples to apples when benchmarking algorithms with each other.[5]

For continuous outcomes (e.g., number of hairs), the Gini and K-S metrics don't make sense (since they are defined for binary outcomes). And even worse, there is no similarly elegant and satisfying metric available. Correlation metrics have a tendency to be unduly *biased* by outliers—single cases that are very different from all other cases (imagine a single gorilla man in your sample who is huge and also, due to some genetic defect, covered in a dense fur). A very practical approach that I have found very useful is to look at the dispersion between deciles. A *decile* is an extremely useful thing—it takes a sample and divides it into ten equally sized buckets (you also can create just four or five buckets—called *quartiles* or *quintiles*—or really any other number of buckets as long as the number of cases per bucket is large enough to be meaningful—remember my rule of thumb of 100-200 cases). You then simply measure the ratio between the average outcome of the decile with the lowest predictions and the decile with the highest predictions.

Example: Let's assume I have applied my hair algorithm on a sample of 1,000 people. The 100 cases with the smallest number of hair predicted range from 47,312 to 63,820 hairs; the 100 cases with the largest number of hair predicted

---

[5]The Basel Committee on Banking Supervision published a very comprehensive review of both common and more arcane metrics for assessing the rank ordering of algorithms in its Working Paper No. 14 (May 2005) titled "Studies on the Validation of Internal Rating Systems." It singles out Gini and K-S as most useful as well.

range from 153,901 to 178,888 hairs. Now I calculate the average actual number of hairs in each of the two deciles, so I might get 51,123 and 181,309.[6] The ratio of the two is 3.5, which means that in the top decile, the number of hairs is on average 3.5 times larger than in the bottom decile. For all practical purposes, this is a *material* and hence useful difference—my algorithm clearly is helpful in, say, deciding how much to pay for the hair of a given person.

Examining deciles (or whatever number of buckets you choose) is good for many reasons. You also can calculate Gini on binary outcomes on the basis of deciles, and it comes in handy for assessing calibration (see the next section). The notion of a "group" of people with a low or high estimate also is helpful in further root cause analysis—for example, you could decide to "have a look" at people with a very high estimated number of hairs and look up a handful of cases within that bucket or even visit them in person (watch out for gorillas, though!).

Just as in your distribution analysis, the analysis of rank-ordering becomes most meaningful once you benchmark results (e.g., if you know that typically a good credit scorecard for approving loans to small businesses has a Gini of 60-75 (assuming there is a credit bureau in your market) but your bank has a Gini of 30-40, then it is likely that your scorecard has a problem, which may be a bias against small businesses, the issue with small delinquencies reported in the credit bureau discussed above, or a bias of your data scientist against "creative" (but for this segment critical) data sources such as the way how the owner of the small business uses your banking app for his private savings account). It is also hugely informative (and concerning) if you observe a sudden drop in the rank-ordering performance. And it *can* also make sense to drill down into individual segments.

If you divide a population into various segments and calculate the rank-ordering capability of an algorithm by segment, you often find substantial differences. For example, I once had a client where the small-business application model achieved a Gini of 50 for one segment—but only 12 for another segment! A Gini of 50 is not the best I have seen but definitely sufficient to run a profitable lending business. A Gini of 12 is the equivalent of driving a car after having glued a newspaper to the windshield. If our gorilla is about to step on your car, you will probably see the shadow of his foot through the newspaper—but otherwise you are pretty much blind. (In case this was not blunt enough: *Do not use an algorithm with a Gini of 12 and believe you do much better than rolling a dice!*) The flaw of that model was an inappropriate handling of missing values. The model used various inputs such as the business

---

[6]Have you noticed that the average of the decile with the largest predictions is larger than the single largest prediction made by the algorithm? Well done! This happens in particular with outliers—it looks like our gorilla man was correctly included in the top decile but on an absolute level, the algorithm still has vastly underestimated the number of hairs, treating the furry beast as a human. This is a common phenomenon.

profile of the applicant and the credit bureau report. For some applicants, there was no credit bureau report because the company never used a loan before. In these cases, the model made a very blunt assumption—namely that the applicant had a mediocre, albeit not lethally bad, credit history. This biased approach discriminated against many excellent businesses and greatly harmed the bank's ability to lend to good companies. Happily, our simple analysis observed this bias, and the client solved the problem by creating a new, unbiased algorithm for applicants without a credit history.

However, it is important to stress that diving into subsegments only makes sense to a point. The reason is that Gini measures the *discriminatory power* of an algorithm—which depends on a *dispersion* of outcomes in the underlying data. Let's assume I built an algorithm that tells you whether a person ever has committed a crime. What Gini would you expect if I test it on a population of high security prison inmates? It would be zero! Why? Take the decile of prison inmates with the lowest probability of having committed a crime— how many of them would you expect to actually have committed a crime? Obviously 100%, maybe minus one or two odd cases of miscarriage of justice. The decile with the highest probability of having committed a crime likewise is 100%—so there is no dispersion of the propensity of being a criminal, and therefore it's impossible for the algorithm to show any discriminatory power.

A real-life illustration of this point is a client of mine who implemented a new credit scorecard I built for him. I had estimated that his credit losses would drop by over 40% thanks to the new algorithm, and he was very excited that this was exactly what had happened. However, he then calculated the Gini for my algorithm using a recent vintage of loans—and was greatly upset that the Gini had fallen by 15 points relative to the Gini the algorithm achieved at the time I built it. The explanation was simple: by eliminating the worst applicants (accounting for 40% of all losses), the remaining applicants that got approved by my algorithm were a lot more homogenous and hence *within* that group, the remaining discriminatory power was a lot lower.

In fact, going back to our deciles can illustrate this: imagine for a second all ten deciles. The lowest risk decile might have a default rate of just 0.2%; the highest risk decile might have a default rate of 43%. If I decide to reject the worst decile of applicants, I now have only nine deciles left; the now "worst" decile might have a default rate of just 12%, and if I reject that decile as well, the next worst decile might even have a default rate of just 5%. If I only approve the very best decile, my Gini very well might be close to 0 for this very safe segment.

For this reason, a low Gini of subsegments is merely a useful warning sign but not a certain proof of a problem.

Bonus remark: As already indicated in Chapter 12, I am a lot less fond of **hit ratios** and the calculation of **Type I/II error tables**. These confound the rank-ordering capability of the algorithm with the appropriateness of the entirely independent decision where to draw the cutoff for accept/reject decisions. However, if you only care about whether in the past the correct decisions have been made (based on the algorithm and whatever cutoff your organization has chosen), then these metrics are useful.

How to Assess Calibration of an Algorithm

Now that you are familiar with the concept of bucketing (e.g., in deciles), assessing the calibration of an algorithm becomes a breeze. The key concept to remember is to differentiate calibration issues across the board from calibration issues of individual buckets.

On the highest level, you can compare outcomes with predictions for the entire database at your disposal. Similar to distribution analysis, I strongly recommend you focus on flow numbers—for example, individual vintages of loans. In fact, the wonderful concept of "vintage" deserves a little detour. The concept is also used for expensive French wine, presumably because it's as delectable as the little statistical analysis I'm describing here. Vintage analysis is an approach to create cohorts that enable the comparison of apples with apples (and grapes with grapes). The idea is to lump together *only* cases that were decided in the same time period—the period could be a day, a week, a calendar month, a quarter, or an entire year—and when comparing outcomes between vintages, to always look at outcomes in the same *time period* after origination. So for marketing emails, it would make sense to look at outcomes from today's emails two or three weeks later; for mortgages, you might look at yearly vintages and their default rates five years later.

What does this give you? Track multiple vintages over time and look for a consistent trend. For example, if you track daily click-through rates of an ad and measure what percent of people clicking the ad actually buy your product, you might see that a smaller and smaller percentage of people clicking the ad actually buy it. This might be due to competitors offering a better deal—or your algorithm having developed a harmful bias and thus targeting the wrong people. Similarly, if you track quarterly loan originations, you may find that in the first eight quarters after a loan is given, an ever larger share of customers default. This is an alarming sign and was in fact a canary in the coal mine for the financial industry, where as early as 2005 a vintage analysis of toxic home equity loans (those where the loans were much larger than the value of the house) would have indicated an explosion in defaults. Sadly, vintage analysis is widely underused!

Now let's look at vintages in the context of calibration. Based on a policy to only approve very safe customers (maybe only the first five deciles) you might have expected a default rate of just 1.2% for the loans originated in the first quarter of last year. Yet, to your dismay, you realize that 2.7% of loans have defaulted.

Thanks to the size of your portfolio, you may be able to now again cut your portfolio in deciles and compare the average predicted default rate with the actual default rate of a given vintage *in each decile*. You might find that across the board, for each decile the actual default rates are substantially higher than predicted.[7] This is a manifestation of the algorithm's stability bias—it seems to have been calibrated towards a period with a generally lower default rate than what your portfolio experienced in the last year (maybe because the economy has entered a deep recession or because a lot of customers have developed an unprecedented habit to gamble (and lose money) with cryptocurrencies). The solution to this is a *recalibration* of the algorithm—for example, if your algorithm has a simple linear form of the kind you saw in Chapter 3, your data scientist can recalibrate the algorithm simply by adjusting the constant term c.[8]

On the other hand, if you find that for each decile, the calibration of the algorithm has been spot on except for the fifth decile, where the actual default rate spiked through the roof, then something very different must be going on. The problem clearly must be caused by a small subgroup of applicants who all or mostly seem to end up in a particular credit score range (which is captured by the fifth decile). Now it is time to call in Sherlock Holmes—which we will swiftly do in the next section.

Before we get there, two final points on assessing calibration. First of all, statisticians of course have developed more or less sophisticated approaches to assess the *significance* of any perceived calibration issue. The most frequently used approach is the chi-square test, which basically tells you how likely it is to observe the empirically observed distribution of outcomes across buckets (e.g., deciles) assuming that the algorithm's calibration is correct. The challenge with these statistical tests is that they tend to ring the alarm bell a bit too often. I therefore encourage users to always firmly keep the concept of *materiality* in mind and to focus on situations where calibration issues appear to be both significant and material.

---

[7] It is extremely unlikely that you will observe exactly the same ratio (i.e., 2.7 : 1) in each decile. This is due to two reasons. First, you may realize that rates are bounded between 0% and 100%, which means that any rate of 50% or higher technically never can double. (What actually doubles is the odds ratio, which is an unbounded transformation of rates.) Second, empirically the kind of macroeconomic forces I am alluding to in this example have only a limited effect on very safe borrowers who therefore show less fluctuations across the business cycle than riskier borrowers. As a result, if the population rate rises by a certain factor, very safe and very risky customers tend to see a smaller relative increase in default rates than customers with intermediate risk levels.

[8] Chapter 3 showed the structural form of a linear regression to estimate continuous variables. For binary outcomes, the equivalent structural form is a logistic regression, which also has a constant term. If the algorithm has a more complicated structure, it may not have an explicit constant term but if you convert its output into a logit score and then wrap a logistic function around it, you can make a virtual adjustment of the "implicit" constant term.

Second, it is very tempting to look at an individual case and to assess whether the estimate is reasonable. With the benefit of hindsight (knowing that a customer has not defaulted), it is always easy to say: "How ridiculous that this algorithm predicted a 50% default rate for this company. As we all can see, this customer has been stellar and should have been assigned a probability of default of less than 0.1%" (which would put it into the coveted AAA region). The fallacy is that there might have been another customer who looked similarly safe but who, in fact, has gone up in flames due to an audacious bet on cryptocurrencies. If that customer had been assigned a probability of default of 50% as well, the model actually would have said: "Here are two similarly risky customers of which I expect one to default but I don't know which one." Picking the one who had not defaulted and claiming that the model was wrong is therefore a bit disingenuous. More broadly speaking, even if a more careful analysis (e.g., assessing the default rate of a large group of customers with a very high predicted probability of default) would prove beyond doubt that the algorithm is biased, discussing a single case is not very fair and useful because the speaker is bound to have numerous biases as well (not least the hindsight bias).

Significance and Normal Ranges

As mentioned, to make monitoring of algorithms both practical and efficient (thus ensuring that it actually happens), it is best to define a set of standard reports that are produced periodically (depending on how long it takes to accumulate enough cases in the flow to produce *significant* results—my rule of thumb of at least 100—, I generally recommend this on a weekly to quarterly basis) and to define automated "trip wires" that alert you of a metric being outside of the range you would consider "normal."

I used to have a wise friend (who unfortunately passed away a long time ago) who often would ask me jokingly, "What is normal?" It is important to adjust the guidelines here to really fit your requirements. That means that you should relax the trip wires at least temporarily if you get more alerts than you can handle and end up concluding most of the time "I guess it must be OK." And you should tighten the limits if even deviations from expected values much smaller than the mechanistic guidelines suggested here would feel discomforting or even unacceptable to you. This recognizes that there is no such thing as perfect truth—much in the world is uncertain, and human life is a quest for optimization under uncertainty as well as constant constraints of time and resources.

Having said that, here is a pragmatic approach to setting trip wires:

- **Distribution analysis**: For *single metrics*, the best approach is to use a t-test to compare the distribution in the most recent period with your reference data (e.g., the data you used to develop the algorithm—which may have gone through a lot of scrutiny to avoid or eliminate biases—or a benchmark from the general population). A very pragmatic alternative is for your data scientist friend to make some static assumptions about the standard deviation of more recent data and based on that define a fixed "acceptance range" outside of which deviations in the mean of the tracked distributions are deemed to fail the t-test.[9] And finally, do ask yourself if whatever threshold comes out to be statistically significant also would feel *material* to you and other stakeholders; if not, you may want to increase the range until it also reaches a materiality threshold. By contrast, for tracking the distribution of cases among multiple categories, you can use either the chi-square test or the Population Stability Index (PSI), which is frequently used by banks.[10]

- **Override analysis**: I find it most practical to ground this in a historic benchmark that you may want to adjust for stated strategic objectives. In the absence of any history, I generally find override rates above 20% warranting some investigation. The truth is that for a high override rate you must believe that human decision-makers are substantially better than the algorithm and that the importance of the decision also warrants the time humans spend on an override—and empirically that's a high bar for most humans to meet. On the other hand, if you know that your algorithm has certain blind spots and your concern is that "humans in the loop" do not spend enough time challenging the algorithm, you may also want to specify a minimum override rate.

---

[9]Assume that in your reference data set 50% of the population are female but that only 20% of applicants approved by your CV screening algorithm are women. As we discussed, whether or not 20% versus 50% is a *significant* difference will depend also on the absolute number of applicants approved by the algorithm in the period under consideration. This pragmatic alternative to a t-test does not consider the actual number of cases but simply assumes something—for example, that each quarter you are screening at least 100 applicants and therefore 20% would be a significant problem.

[10]PSI actually is proportional to chi-square, as has been shown by Bilal Yurdakul in his dissertation, "Statistical Properties of Population Stability Index" (2018).

- **Calibration analysis**: While the chi-square test does allow you to flag outcomes that deviate from expected values by a certain confidence level, my pragmatic recommendation would be to define thresholds by materiality instead. For example, if your algorithm predicted a probability of default of 2.5% but economically you wouldn't be fazed as long as default rates are below 3%, then 3% is a good benchmark. A 20% increase in the rate of something generally is not a particularly big variation (for comparison, during financial crises, I have seen default rates rising to levels of five or six times the average of "normal" times).

- **Rank-ordering analysis**: For Gini (binary outcomes), you can define benchmarks based on the development sample and external benchmarks minus a materiality threshold. Assuming you have not worked much with Gini before, a tolerance level of 5 Gini points might be a good starting point. The smaller the case quantity, the longer the time periods you consider, and the more volatile the context, the more easily an algorithm's Gini will drop by more than 5 points.

If you feel unsure about your triggers, another approach is that you first decide how many issues you can afford to investigate in a given period, and then for each analysis, you look at the biggest deviations to decide where to draw the "cutoff" in order to get the desired number of warning signals. For example, if you decide that for each analysis, you would like to initially review about two situations, check for each analysis what the second-largest deviation is and set the trigger for that analysis somewhere between the second and third largest deviation. The good news about this approach is that it is both manageable and actually forces you to spend some time with your algorithms and the data. If any root cause analysis identifies a problem, naturally you'll ask yourself if other algorithms or subsegments might suffer from the same issue and expand your investigations accordingly.

In the next section, we will discuss what you should do if a particular metric triggers the trip wire.

Root Cause Analysis

We have discussed four broad lines of attack to detect algorithmic bias: distribution analysis, override analysis, and the assessment of calibration and rank-ordering. All four approaches work a bit like a motion sensor—if an alarm goes off, we know that something has moved in the garden but we don't know yet whether it was a burglar or the neighbor's cat.

One can take a positive or a negative view on root cause analysis. The negative view would lament that there is no easy silver bullet to immediately catch the culprit; the positive view is that this is actually fun because finding the root cause of algorithmic bias is not at all about dry statistics but about discovering a hidden story that is full of the many incredible quirks of life. Data wants to talk to you—in fact, it is *dying* to tell you its story –you just need to ask the right questions and bring some patience because sometimes the story data will tell you is convoluted. To be honest, my Mom is quite the same!

Much of the detective work you will need to do is to trace the roots of an anomaly in the algorithmic output to an underlying cause—maybe a problem in the input data or an issue in the process generating the data. There are two tools that are central in this endeavor:

- I always love to use distribution analyses not just of outputs but also inputs. For example, if there are 12 inputs in the algorithm and the calibration analysis suggested that there is a particular class of cases for which the algorithm systematically is off, compare the distribution of the 12 inputs between the development sample and the period in which the issue popped up. Maybe it is the category "other" in one of the 12 inputs that suddenly has changed its meaning, so suddenly having a lot more or less cases in "other" would drop a hint and thus give you a shortcut to what to look for in the real world (as you now will have to talk to people to find out what kind of cases moved in or out of the category "other").

- The other tool is the decision tree (e.g., a chi-square automatic interaction detector, commonly referred to as CHAID) to identify factors that narrow down the subsegment where the issue is—a tool I mentioned above. Once you have defined what you are looking for (e.g., cases with an override or cases with a big classification error as measured by the difference between predicted and actual value), run a decision tree. The art is to decide how many and which predictive variables to provide. If you have thousands of predictive features and provide all of them to the decision tree, you very well may strike gold (i.e., the decision tree tells you loud and clear where the problem is)—or maybe not. Let's assume you had a sudden spike in defaults and the first try (throwing all your 5,000+ features into the algorithm finding the tree) suggests customers who have not connected to the bank through any channel for two weeks have a higher risk of default. Rings a bell? If not, try a decision tree using only geographic markers as predictors. Bingo— you might realize that the spike in defaults is linked to a coastal region struck by a hurricane last year; the natural disaster destroyed many homes as well as the commercial infrastructure and thus caused many defaults and also interrupted communications between the bank and its customers for two weeks. (This is an example of what we called a *traumatic event* before—it's a one-time phenomenon in your (data) history that can seriously bias your algorithm and requires forceful intervention by a data scientist. It also illustrates why it's always good to talk with other people about your root cause analysis— maybe someone who was more closely involved in the disaster relief work would remember people not being able to connect with their banks and hence even would get the enigmatic hint your first decision tree gave.)

A systematic approach to test for every possible root cause would be a prohibitive effort—consultants often call this *boiling the ocean* (not a very efficient way to find a couple of shrimp for lunch, let alone the environmental damage). Instead, it is best to follow a hypothesis-driven approach: make an educated guess about what most likely has happened based on everything you know about the data (including the basic monitoring but also the genesis of the algorithm) and the underlying real-life situation, and then rigorously test your hypothesis. If the data doesn't support it, think of the second-best hypothesis, and so on. This is the fun part—just think how Sherlock Holmes would inspect the situation, find clues, and let his imagination pair up with his broad knowledge of many disciplines to solve mysteries.

To aid this process (and in line with our hypothesis-driven approach), in the following I will briefly revisit the major biases discussed in Part II of this book and suggest how they may show up in our basic reports. This will tell you where to look if you want to investigate a particular hypothesis.

- **Missing rows of data**: Be it because the Bavarian data scientist was biased and didn't include Belgian beers in the database out of conviction (admittedly, beers like the Mort Subite Oude Kriek, made with real cherries, really do not comply with the German Beer Purity Law) or because of an honest mistake (maybe he filtered beer from other beverages using a free-text data field called "beverage category" and missed that beer produced in the French-speaking parts of Belgium is labeled "bière" rather than "bier," which is the Dutch version), missing rows of data usually reveal themselves if you look at summary statistics: how many cases do you have in your database and what is the total (e.g., monetary value, liters, or bottles), and how do these totals compare with external statistics? I am always amazed how often the initial dataset data scientists receive from someone else is incomplete, and how often data scientists lack the external benchmarks to easily check whether the dataset actually appears to be complete. And be creative: if you don't have the total beer consumption of Belgium as a benchmark, calculate the average consumption per capita according to your dataset—that's a number you can benchmark with other countries.

- **Missing columns (i.e., missing features/predictive variables)**: This one hinges on your practical content knowledge. Consultants often speak about the "front line"—the workers actually screwing together stuff in factories or selling stuff to customers or repairing stuff in body shops. These guys know what *really* happens (including what goes into a sausage)—whereas people dwelling in company headquarters, academia, and analytics centers often have what I call *textbook knowledge*—an idealized view of the world more grounded in what ideally *should* happen. If you haven't done so yet, ask the front-line people what *they* think drives outcomes, and check how much of that is included in the dataset.

- **Subjective data**: If the distribution of specific input parameters (the features used by the model) is different from real life, often the assessment of such a parameter is compromised. If you don't have any benchmark distribution to compare with, see if the data looks normal in the statistical sense. "Medium" values should be frequent, extreme values rare; for amounts (which can become very large—think income—but have a limit at the lower end, e.g., due to minimum wage or at least because they can't become negative), expect a lognormal distribution, which means that the logarithm of the amount should be normally distributed. Hint: If 80% of all companies applying for a loan in your bank are deemed to have "exceptionally good management," this would *not* be normal (even if the companies in question are Australian)!

- **Traumatizing events reflected in the data**: This one usually reveals itself in unusual spikes in the outcome variable (e.g., the majority of accounts in a branch or region having defaulted in a particular month); often this also leads to violations in the rank-ordering of the algorithm (e.g., a lower-risk decile has a higher default rate than the next-risky decile). Decision trees are amazing at localizing such pockets but it requires context knowledge to connect the dots and recognize the real-life event mirrored in the data.

- **Outliers**: These little devils are often hiding in the details. The distribution analysis will only indicate outliers if it is designed to call out the minimum and maximum value observed in the data—which therefore is critical. And given that one work step in the model development is the *treatment* of outliers (where outliers often get "cut back" to a value range considered "normal," which is intended to make the outlier less influential on the model equation), the question arises whether you should look at distributions of input values before or after outlier treatment. Here my advice is to look at distributions *before* treatment—the reason is that the outlier treatment may turn out to be insufficient or inadequate, so I much rather have a treated outlier still visible and thus triggering a discussion about the treatment than an invisible but poorly treated outlier hiding in a "normal" range and causing bias. If the same type of outlier comes up again and again and you are confident that the treatment is OK, you may decide to move to monitoring treated values for that particular variable.

- **Artifacts in the data created by inappropriate data cleaning**: For this one, you obviously need to analyze features *after* data cleaning; the feature affected then can often be identified by both the distribution analysis (as the treatment might have caused a new concentration in a particular value, such as zero) and a decision tree (including a tree isolating cases with high overrides); some cases also show up in issues with rank-ordering or calibration (because for the affected cases the algorithm wouldn't work properly).

- **Stability bias**: If the time period used for developing the algorithm is structurally different from the time period in which the algorithm subsequently was deployed, usually calibration is off across the board and rank-ordering performance drops.

- **Dynamic development of biases through user interaction**: Ironically, this bias can lead to an *increase* in rank-ordering performance over time (it is like user and algorithm entering a dance in lock-step); a distribution analysis of outcomes (as well as predictions) also might show an increasing concentration in a few items. Most importantly, however, you need to discuss how the algorithm is used and develop pattern recognition if the way the algorithm is updated allows for harmful feedback loops.

- **Real world bias**: Here your distribution analysis would show a bias that is mirrored by a similarly biased distribution in external benchmarks (e.g., your CV screening algorithm comes up with a distribution heavily skewed towards men that is mirrored by the distribution between men and women in your entire organization).

When you have to literally crack a "tough" walnut with metal tongs, you often have to try a couple of different angles until the walnut yields to the tongs. This exploratory data analysis is similar—do not expect your first analysis to always give away your data's secret! For example, when you suspect a real-world bias, you might be surprised to find a simple distribution analysis to suggest the opposite: Martians may have a *higher* approval rate at top Zeta Reticulan colleges than a Zeta Reticulan. This may be due to an implicit self-selection bias, however: if most Martians don't even bother to apply at these colleges because they expect to be discriminated against, the few Martians who do apply probably are stand-outs that may also get special encouragement. In this case, only comparing the distribution of *applicants* (as opposed to *approval rates*) between Martians and Zeta Reticulans with the overall proportions of society will reveal what is really going on.

And of course there are many more possible techniques to identify root causes. This book isn't meant to be a data science university, and your data scientist friend certainly will be most helpful in suggesting alternative or additional analyses you could run to get to the bottom of a particular issue. This is also an area of continuous innovation. On the one hand, there is more and more visualization software for data, and one should never underestimate the value of *looking* at data (e.g., in order to visually detect outliers or other odd cases—in this task, our brain is really at its best because this is a key way nature detects threats); on the other hand, machine learning especially has enabled very advanced new tools such as *anomaly detection* to flag potential root causes for biases (or trouble, more broadly).

Finally, when you do your root cause analysis, you may sometimes find multiple leads and wonder which ones are really material. If you are concerned that imperfections of a particular feature might drive outcome bias but are unsure about the materiality of the effect, you can assess the materiality of such a feature by putting in the median[11] (continuous variables) or median-equivalent (categorical variables)[12] for all other variables used by the algorithm (thus creating a synthetic "typical" or "average" case) and then simulating how much a change in the variable in question actually drives outcomes (as would become evident by comparing the average prediction or, for a more elaborate assessment, through a comparative distribution analysis of outcomes for alternative value choices). This approach is blunt (e.g., it disregards possible interaction effects where your variable is affecting outcomes only in a certain subgroup of cases, which may be different from the "average" case you are exploring) but often enough sufficient.

We now have completed the arc from regularly monitoring an algorithm by tracking a handful of basic metrics to completing a root cause analysis in order to understand why a particular metric in the monitoring report has rung a warning signal. However, so far we implicitly have focused on the easiest context: relatively simple algorithms where we know the limited number of input factors. Unfortunately, many machine learning algorithms are somewhat more challenging to monitor. We therefore will briefly discuss what is special about them—namely a higher degree of complexity (which is why machine learning algorithms often are called a "black box") and potentially a continuous, automated updating of the algorithm by itself.

Monitoring "Black Box" Algorithms

"Black box" algorithms are predictive models developed by machine learning that may have hundreds or thousands of input variables and where the mechanics of the algorithm are too complex for a human to review and understand in detail. They still assign an estimate to every case, so the basic

---

[11]The median is similar to average except that it is the actual value of a case where 50% of the sample are lower and 50% are higher (hence the case is "in the middle"). Assume you are in a country where 33 million people earn $1,000 per month each, 33 million people earn $2,000, 33 million people earn $3,000, and 100 tycoons earn $1 *billion* per month each. The average income in this country is $3,010 but the median is only $2,000—the median therefore is a much better proxy for the "typical" income and what the average income of "normal" people is.

[12]While for continuous variables (e.g., income) it is easy to calculate the average or median, this is impossible for categorical variables (e.g., job—you obviously never want to say that a tax collector is the average of an accountant and a robber!). Also, the "mode" (i.e., the most frequent category) often is a poor proxy for the average because often the most frequent category is at the lower end (e.g., the most common jobs often are relatively poorly paid). A better approach is to sort categories by their median outcome (e.g., median income of each job category), then identify the "median" income, and look up which category (i.e., here job) is closest to the population median (e.g., computer programmers).

monitoring works like with any other statistical model. The difficulties arise when you attempt a root cause analysis. In particular, you will face three challenges:

- You will deal with a much larger number of factors going into the model that could each have caused a bias detected by the monitoring, so you may feel like you're searching for a needle in a haystack.

- The development algorithm automatically does much of what the data scientist traditionally has done manually: treat outliers and missing values or code interaction effects. As a result, there are a lot more ways a machine learning model could have inadvertently caused bias but nobody with whom to discuss and challenge the treatment approach.

- You will find it much harder to develop hypotheses on what might have caused a bias because you neither know the mechanics of the model nor is your data scientist friend likely to have become as intimately familiar with the data as in the past when she had to do a lot more manual work to fit a model.

To overcome these challenges, there has been a push to develop so-called *XAI* (explainable artificial intelligence). There are three elements to it:

- Your data scientist may or may not have chosen machine learning techniques that are *relatively* more transparent than others. These choices are discussed in Part IV and are outside of the scope of this chapter.

- The **globally most important drivers** of a machine learning model and the direction of their influence on outcomes can be visualized by a process called *perturbation*—your data scientist friend essentially simulates how changes (generated through random noise) in the various input factors change the model's outputs and then mechanically searches for the factors accomplishing the greatest variations.

- **Locally most important drivers** refer to a particular case (i.e., an applicant that seems to have been unfairly discriminated against by a bias); here the perturbation analysis indicates what factors had the greatest impact on *this* case.

If your basic monitoring has raised a concern with a machine learning model, the first step is to ask your data scientist for an analysis indicating the globally most important drivers. There is no hard and fast rule of where to draw the line between "important" and "not important" but keeping the idea of materiality in mind, you will probably end up focusing initially on not more than 5-12 variables. You can mechanically work your way from the single most important driver down the list, or you can use your business judgment to prioritize factors where you sense the greatest likelihood of a bias.

Decision trees—the clever tool we found in our root cause analysis toolbox—also work with hundreds or thousands of model inputs. In fact, many machine learning models are so-called forests of hundreds of trees—but the overload of variables might cause funny effects that make it hard to interpret them. If this is a problem, you should ask your data scientist to produce a Principal Component Analysis (PCA) of all the factors, conduct a VARIMAX rotation (a rearrangement of the numbers that make them a great deal easier to interpret), and point out for each of the main components the most highly correlated variables.

The PCA is a little miracle that recognizes that many variables are quite correlated (i.e., measure essentially the same thing). For example, a lot of attributes of an applicant may all represent income. Another bunch of variables may indicate conscientiousness (the personality trait that drives risk; highly conscientious people plan ahead, are circumspect, and have excellent impulse control to keep themselves out of trouble), and so forth. The PCA looks at all your data and then comes back saying, "I see that you have 5,000-odd factors; actually, most of the data is about five big themes, and the first theme seems to be all about income, the second theme about conscientiousness, and so on." And you then can pick for each theme one variable (e.g., your most reliable metric of income and the best metric you have on conscientiousness) and build a decision tree with this very short list of variables.

As you might have guessed, doing a PCA in practice is slightly more difficult than the previous paragraph made it sound like—there are in particular challenges to be overcome if you have missing values or categorical factors—but that is why data scientists spend so much time at university! You will have to discuss with your data scientist what is practically possible for a particular dataset and may have to settle for something slightly simpler but fundamentally what you need and should be able to get is a much more condensed list of factors where redundant variables have been eliminated. (Sadly this approach is not taught in every data science course, so it may be novel to some data scientists—stick to your guns on this one! I discuss it in detail in Chapter 19.)

Through all of this work you may end up zeroing in on a particular group of cases where the problem seems to be centered. This is the moment where an analysis of the *locally* most important drivers can come in. For example, if your basic monitoring raised a suspicion that your credit application model discriminates against certain Martians, in your initial root cause analysis a decision tree tasked to find hot spots amongst overrides or wrong predictions (especially rejections of applicants who in hindsight have repaid all of their debts just fine) might have identified that they are all living in a handful of ZIP codes known to be Martian ghettos.

In the US, such redlining would be illegal. Maybe it is illegal in Zeta Reticuli as well, and your data scientist points out that actually ZIP codes are NOT provided to the machine learning model as an input! Incredulously, you ask for an analysis of the locally most important drivers for a sample of the Martians who live in these ghettos. You find that most of the normal drivers of risk (e.g., income and the credit bureau history) are conspicuously absent from the list of variables you receive from your data scientist. Instead, the single most important driver is the distance of the applicant's home to the nearest branch of Joe's Potato, a famous potato fast-food chain, which obviously explains the mystery![13]

This example also shows a big challenge with machine learning algorithms when it comes to biases: if the bias is already present in the data used to train the model (e.g., because it mirrors societal biases), the machine learning algorithm will go out of its way to capture indicators for the bias. If you remove direct indicators (e.g., the ZIP code), it will find indirect ones (e.g., the distance to Joe's Potato). If you remove the indirect ones, it will find even more indirect ones (e.g., the number of businesses with a name starting with "J" in the vicinity of the applicant). This is why it is not always possible to remove the bias from the model and we also must consider alternative solutions (which we will cover in the subsequent chapters).

Monitoring Self-Improving Algorithms

Because machine learning is so fast, it has enabled another innovation in the field of analytics: self-improving algorithms. Traditionally, a data scientist collected data, got to know the data through various exploratory analyses, and then crafted a predictive formula through a number of iterations where she created features, corrected data issues, and nudged the statistical algorithm in a sensible direction through the choice of variables and hyperparameters. Machine learning can do all of this (minus the "sensible") automatically—and thus can churn out a new version of a model every week, day, or (possibly) even minute.

---

[13]I am sure you know that the favorite food of Martians is potatoes—so Joe's Potato naturally has most of its branches in areas with a mostly Martian population, and the five Martian ghettos you identified account for a whopping 60% of all branches of Joe's Potato. This distance variable therefore created a backdoor to redlining for the algorithm.

The challenge for our oversight is thus that by the time we have analyzed a machine learning model and found the root cause of a bias (or maybe gladly established that there is no bias), the machine has already created five more versions of the model that now may look totally different from the model we examined. How on Earth (or Zeta Reticuli, for that matter) can we keep up?

The additional step required is to also establish a monitoring for each new version of the algorithm to establish when it is "materially" different from the last version we examined and explicitly approved for use. I think of this like hiring a person: when we interview job candidates, we try to figure out how apt they would be at our job; if we hire a cook, we might even ask her to demonstrate her skills in our kitchen. Of course, once the cook has started to work for us, things might change—we might decide to put guinea pig on the menu (known as cuy, this is a specialty in the Peruvian Andean region, although I must warn you that the commercial success of this menu choice outside of Peru might be a bit doubtful) and realize that our cook is not well trained in the preparation of guinea pig, or our cook might develop a bit of an alcohol problem and our customers might start complaining that our zabaglione tastes a lot worse since most of the Marsala wine ends up in our cook's stomach instead of the zabaglione. We therefore need to regularly monitor our staff, and when warning signals (e.g., a drop in our TripAdvisor ratings or staggering cooks in our kitchen) occur, we need to investigate, possibly putting a staff member on leave until we are satisfied that the person is no threat to our business.

Applying this thinking to machine learning, self-improving models therefore also should auto-produce a list of metrics that measure the degree to which a new version of a model is different from the last version of the model that we explicitly tested and approved. For example, the script updating the model could calculate predictions for the same set of applicants with both the old and new model, and calculate the degree to which predictions are different—if changes are too dramatic, a stop signal would be triggered. Or the algorithm could identify the globally most important drivers and compare them with the previous version.

In Chapter 22, I will discuss in greater detail how best to monitor self-improving machine learning algorithms. For now, let me just stress how critical it is that business users discuss with data scientists how much change they are willing to accept (there is no one right answer—it really is a tradeoff between the business upside from rapidly updating decision algorithms and the appetite for risk—very often, for example in fraud analytics, time really *is* money), and that data scientists ensure that a reliable mechanism is in place to put the deployment of a self-updated model version on hold if it seems to deviate too much from earlier versions and hence manual, off-line testing should be completed before the new version is allowed to go live.

Summary

In this shortish chapter, we discussed how users can monitor algorithms to detect bias without being a data scientist themselves. Key recommendations are:

- A basic monitoring regime should include both forward-looking metrics (which can be calculated even before predicted outcomes have materialized) and backward-looking metrics (which compare predictions with actual outcomes).

- Two very useful forward-looking metrics are the distribution analysis and the override analysis.

- Two very useful backward-looking metrics are the calibration analysis and the rank-ordering analysis.

- Distributions can be described by the rate of a binary outcome happening, the average of a continuous value, or the (percent) distribution of cases between different types.

- Analyzing the flow of cases usually produces a lot more meaningful metrics than the stock of cases, especially for distribution analyses.

- Differences in distributions should be filtered for significance and materiality to avoid false alarms.

- *Marginal* significance can test whether a particular attribute affects an outcome once differences in all other attributes of a case are controlled for and therefore further reduces false alarms.

- If there is a "human in the loop," you may be able to...

  - Compare whether the absolute level of overrides is too high or too low compared to what you deem consistent with a functioning decision process.

  - Identify classes of cases with a particular concentration of overrides ("hot spots").

  - Review the most frequent reasons for overrides to identify biases (or other shortcomings) of the algorithms.

- Assessing calibration will indicate whether an algorithm's predictions are right on average (and for specific buckets, such as deciles) or biased.

- Assessing rank ordering will indicate how powerful the algorithm is in differentiating cases with very different expected outcomes (and where a bias might have impaired the algorithm's rank-ordering ability).

- While for binary outcomes, the Gini coefficient and the K-S statistic are very elegant and useful metrics of rank-ordering performance, assessing rank ordering for continuous outputs is clumsier. As a pragmatic solution, I recommend measuring a multiple that compares the average outcome of the lowest-ranked and the highest-ranked decile according to the algorithm.

- For each metric in your basic monitoring report, you should define a "normal" range so that deviations that merit further investigation can be automatically flagged. These ranges can be informed by significance tests as well as external benchmarks.

- If a metric rings an alarm bell, you should conduct a root cause analysis to see whether a bias is present.

- A root cause analysis often will involve distribution analyses also of input variables, decision trees to profile hot spots, and discussions with front-line people to better understand what is going on in real life.

- You will be most efficient in your root cause analysis if you can develop hypotheses about which biases are most likely to be present and then target your root cause analyses at testing these particular hypotheses.

- "Black box" models developed by machine learning can be monitored in the same way as simpler algorithms but complicate the root cause analysis. A technique called perturbation allows you to identify both the *globally* and *locally* most important drivers as tools to aid and simplify the root cause analysis for such models.

- And if machine learning automatically updates algorithms, it is important to also automatically monitor the degree to which these models change and to suspend the implementation of a new model if it deviates too much from the last validated and approved model (and hence constitutes too high a risk of bias given the risk appetite and context of the business owner).

Identifying the root cause may inform ways to address biases in the algorithm; however, not all biases (especially those rooted in the real world) can be removed from an algorithm. In the next chapter, we will discuss managerial strategies for dealing with algorithmic biases, and after that we will discuss how to create new data that is free of biases.

Managerial Strategies for Correcting Algorithmic Bias

I still remember my shock when I interviewed a Chief Risk Officer of a bank about the key risk drivers she considered when underwriting credit, and one of the first things she said was that homosexuals are "obviously" risky (and hence should be avoided as borrowers).

While one part of my brain hotly debated whether I should confront the clearly intelligent and educated woman with all the arguments why I considered this a horrible bias, another part of my brain reflected on what I had read about her country before my trip—specifically that homosexuals in that country faced a high risk of being killed from hate crimes and seldom lived to old age. However terrible the practice, I had to admit that not lending money to a customer who you fear might get killed any day was simply prudent, and doing otherwise would have been gross neglect.

© Tobias Baer 2019
T. Baer, *Understand, Manage, and Prevent Algorithmic Bias,*
https://doi.org/10.1007/978-1-4842-4885-0_16

While it turned out that the bank did not have any data on the sexual orientation of their customers and this therefore ruled out including this attribute in the algorithm we developed for the bank, it definitely made the question of how to deal with an algorithm that reflects a horrible societal bias real and personal. This is a good opportunity to take a step back and consider that algorithms are neither a god nor a boss but simply a tool, and that as users of algorithms, we have degrees of freedom if and how to use them.

It is also a good time to reflect on the fact that we humans have a long history of effecting change on our environment—this is essentially what public policy-making is about. And indeed, many examples exist of how biases have been tackled—some with more, some with less success. Discrimination against female musicians has all but disappeared since orchestras started to conceal the identity and thus gender of auditioning applicants behind curtains. Quotas are frequent (e.g., in university admissions and for jobs ranging from entry-level traineeships to board memberships), although both their effectiveness and appropriateness continue to be debated.

Going back to the discussion of the overall design of a decision algorithm, there definitely are possibilities for confining the use of algorithms to a subordinate role—for example, algorithms could be used to pick applicants within allotments of specific quotas (as opposed to abolishing quotes by allowing algorithms to completely take over selections), or several competing algorithms could be used (each with its own contingent of approvals), especially if it is anything but clear what even constitutes a good or bad outcome. (For example, in Germany, university admissions for studies of medicine allocate spaces by at least three algorithms—while one strictly goes by academic merit, another one also gives credit for how many years an applicant has waited for a spot, and a third is a qualitative assessment through an interview—arguably competing approaches that assess suitability for the medical profession through academic merit, persistence, and a more qualitative perspective on who would make a good doctor.) An appeals process—allowing for judgmental overrides of algorithms—is another common technique.

If you know under what circumstances the algorithm is biased, you also can exclude certain cases from an algorithmic assessment—maybe comparable to the way academic tests such as TOEFL make accommodations for disabled students (e.g., if you have a speech disorder, the speaking section of the test can be omitted in order to prevent the scoring algorithm from being biased against you in its assessment of your command of the English language).

All of these approaches can help to overcome biases in an algorithm while avoiding tinkering with the algorithm itself, and therefore are worth considering if an algorithm reflects a deeper bias in society or the data that cannot easily be purged from the algorithm.

However, is there still an opportunity to fix the algorithm itself? After all, this may have advantages because a "fixed" algorithm can automate and therefore dramatically speed up decisions, and in terms of making change happen, it is much easier to change the computer code in a central decision engine than to train maybe thousands of front-line people to correct a harmful bias through a judgmental override.

There is a precedent of sort half way to changing the algorithm itself, namely overriding inputs such as faulty credit bureau information. However, now we would like to go further and manually tweak the algorithm itself (something like conducting brain surgery on a statistical formula).

As a matter of fact, this can be done. As a first step, we can include the source of discrimination (e.g., race) as an explicit variable in the algorithm—essentially grab the elephant in the room by its tusks to make it visible to all. In a second step, we need to guide the elephant gently out of the door—which we can accomplish by hardwiring the algorithm to treat every single applicant the same. For example, once we have introduced a race variable, we can program the algorithm to assume that *every* applicant is Martian, regardless whether the person is Martian, Zeta Reticulan, or Klingon.

Such a technique is barely heard of in many disciplines, neither in practice nor in theory, and of course arbitrarily changing an algorithm seems to grossly violate statistical orthodoxy. Statisticians might be quick to point out that this would be a horrible thing to do—we are consciously introducing a data error, and given the correlation of our race variable with other attributes used in the algorithm, all sorts of things might go wrong.

Then again, discrimination and societal bias is horrible, too, and sometimes one evil (such as the sting of the syringe delivering a life-saving vaccine) is really a lot smaller than another evil (such as diphtheria or pertussis). And more to the point, of course a data scientist can run analytics to understand for a sample what the proposed tweak will do and thus get some comfort that the adjusted algorithm will still produce sensible outcomes. For example, in a credit score, one can compare the approval rate and the credit losses estimated by the original (biased) scorecard with the outcomes for the adjusted scorecard. It is possible (in fact, likely) that expected loss rates will go up—however, this can be accounted for by adjusting risk-based pricing (i.e., increasing the interest rates charged) or by somewhat reducing the maximum risk level for which loans are approved. In other words, a "bias-free" bank loan might be a bit more expensive similar to how fair-trade coffee comes at a premium price—but both products can be profitable and safe to use.[1]

---

[1] An important implication is that regulations prohibiting certain biases also can increase cost and therefore prices—very few things in life do not come at a cost. Then again, it is also perfectly perceivable that Martians only approved thanks to your little tweak are exceptionally good customers—be it for unexpected intrinsic reasons or because they are so grateful that they are particularly motivated to demonstrate how much they deserve credit.

One also can run additional analyses to better understand the properties of the variable used to correct for the external bias—For example, a correlation analysis can indicate which other factors (e.g., income or education) are related to the adjustment variable (e.g., race). In particular, by suppressing all correlated variables (similar to the use of Principal Components to reduce redundant variables in the dataset), the data scientist can force the "bias" variable to capture the indirect effects (such as income and education effects) as well. If we now ask the algorithm to pretend that all Zeta Reticulans are Martians, the algorithm will "assume" that also income and education will be Martian-like.

As a matter of fact, economists build little equations to "explain" the world and then evaluate alternative scenarios (e.g., a new tax or a new government spending program) all the time. Yes, having a credit score "assume" that a Zeta Reticulan is Martian takes us into a hypothetical world—but once this sleight of hand gently changed the fate of our credit applicants, it is also very possible that reality will willingly follow our nudge.

How altered predictions can create their own reality is neatly pointed out by the famous experiment that school teacher Jane Elliott did with her third-graders in Iowa.[2] On the day after Martin Luther King Jr. was shot, she introduced an arbitrary bias in the class: children with brown eyes, she explained, had more melanin in their bodies, which not only gave them brown eyes but also made them "cleaner and smarter." The results were hair-raising—not only did brown-eyed kids taunt blue-eyed kids and show through their comments that they actually *believed* blue-eyed kids to be inferior but they also displayed more self-confidence (which can have far-reaching mental health and performance benefits) and actually *did* perform better on math tests, while blue-eyed kids—even former stars of the class—suddenly *did* struggle with tests. And when the teacher told the kids a week later that she had gotten it the wrong way and actually brown-eyed kids were *inferior* to and *less intelligent* than lighter-eyed kids, behaviors reversed. Other psychological experiments have confirmed this point: to a surprisingly large extent we are who we think we are, and hence biases others have towards us subtly and subconsciously can shape our behaviors to comply with these prejudices.[3] This implies that once an algorithm stamps an applicant as "worthy," this can have a real impact on how that person behaves (as well as how others treat that person), such that the tweaked algorithm might indeed create a new, better reality.

---

[2]www.smithsonianmag.com/science-nature/lesson-of-a-lifetime-72754306/

[3]Also, job interviews illustrate this point: If an interviewer *believes* that an applicant is ill suited for a job (e.g., because of a bias against males with red finger nails), applicants actually *do* perform worse, apparently because they pick up on the interviewer's beliefs through body language and subconsciously adjust their behavior, which once again proves that humans can be as bad or worse than biased algorithms! Read more about the impact of subconscious biases on interview and workplace performance at www.forbes.com/sites/taraswart/2018/05/21/prejudice-at-work.

It is also worth noting that the degree to which stakeholders are open to "adjusting" algorithms can be a matter of positioning. If an algorithm is used to estimate the probability of a particular outcome (e.g., completion of a college degree or committing another crime), it is positioned to state a "reality," and hence changing that estimate can feel patently wrong. Mathematically, however, these probabilities are often calculated by plugging a so-called score into a formula (e.g., logistic regression estimates a "logit score" s as a linear combination of predictive features; the estimated probability is derived from s as $1 / (1 + \exp(-s))$). Scores are used in modern lives all the time (e.g., we have credit scores, ride-hailing apps calculate scores for both drivers and riders, there are point scores in online games, and sometimes we even might jokingly declare to our friends how they just earned some bonus or penalty points with a (mis)deed), and giving someone a "bonus score" (e.g., to overcome a societal bias or achieve other policy goals) doesn't feel that wrong—even though in the end, it will have exactly the same impact on the algorithmic prediction (and the exact bonus score to be given can be calculated with the approach discussed—if we introduce a 0/1 dummy variable to mark Martians, the coefficient of that variable would indicate the required score adjustment to give Martians equal footing with Zeta Reticulans).

There is no steadfast rule whether or not tampering with algorithms could and should be considered for a particular situation, and I mentioned valid arguments for either way, just as other methods to overcome societal bias (e.g., quotas) are controversial. I do believe, however, that users who are dealing with difficult trade-offs at least should be aware of this option and at times may conclude that doing so might at a minimum warrant an experiment to test real-life consequences.

Summary

In this chapter, you explored how managerial overrides can contain algorithmic bias where statistical methods alone cannot purge the algorithm of harmful biases:

- By adjusting the decision architecture, the wings of biased algorithms can be clipped and decision outcomes constrained, for example, by overlaying quotas or other overrides aimed at reducing or eliminating bias.

- While statistically awkward, there are also valid arguments for adjusting an algorithm itself, including ease of implementation in far-flung organizations and the psychological effect on individuals if an algorithm stamps them "worthy" of something.

- A viable technique to remove strong biases present in the data (e.g., because they mirror societal biases) is to make the bias explicit in the algorithm by introducing indicators for the source of the bias (e.g., an applicant's race) as explanatory variables but then setting these variable to the same value for all cases.

- Simulations can inform the necessary adjustment to decision rules using the algorithm as input, such as increases in loan pricing to cover incremental credit losses.

- And buy-in by diverse stakeholders can become easier to obtain if the adjustment is positioned in a way that feels natural rather than triggering strong objections (e.g., by adjusting generic point scores rather than probabilities for real-world outcomes).

Even better than fixing an algorithm to remove societal bias, of course, would be to find out what would happen if the bias simply did not exist, and use this experience for developing a new algorithm. And sometimes this is possible. In the next chapter, you will learn how to create new, unbiased data through careful experiments that allow the development of unbiased algorithms.

How to Generate Unbiased Data

One motto of our times could be "data is the new gold"—however, it will shine only if it is pure and free of dirt. Biased data can be lethally polluted and thus worthless. For example, a tax authority once asked me to help them build an algorithm to direct customs inspectors to those containers in the port that were most likely to contain contraband. The project could not go ahead because the only data they had was from a very limited number of customs inspections their officers had done in the past year. The problem: Customs inspectors had chosen which containers to check, and they had complete freedom in how to conduct the checks (e.g., they may have limited themselves to opening the first box falling into their hands and accepting the shipment as containing "Louis Vuitton Croisette handbags" because the duffel bags with a "Luis Vitton" label seemed close enough, or they may have completely emptied the container and carefully checked the L's, O's, and T's of the Louis Vuitton stamp of two dozen bags, knowing that variations in these letters are among the frequent tell-tale signs of fake bags).

What would have happened if I had used this data? The thorough inspectors obviously had found a lot more contraband, so they had introduced a strong

© Tobias Baer 2019

T. Baer, *Understand, Manage, and Prevent Algorithmic Bias*,
https://doi.org/10.1007/978-1-4842-4885-0_17

bias in the data—and in attempting to predict where to find contraband, the algorithm actually would have told me which containers historically would have been most likely to be checked by "good" inspectors. Containers with contraband that slipped through because of a half-hearted inspection effort, by contrast, would appear to the algorithm to be OK, and hence in the future it would steer inspectors away from such containers.

Smugglers, of course, are aware that they have better chances of slipping through customs in some ports (maybe even during particular shifts) and are hence prone to directing their shipments accordingly. My biased algorithm now would directly play into their hands by suggesting to reduce checks in these ports and during these shifts even more (because historically they didn't yield much) and instead focus even more resources on the ports that historically were well checked. As a result, I strongly advised the customs authority *against* building an algorithm with the available data.

So what can we do? The solution is to generate new, unbiased data. This is a lot easier said than done. In particular, a process to collect such data would need to meet two important requirements:

- The sample must be random—a container I deem to be most unlikely to contain contraband must be as likely to be checked as a container I am certain to contain contraband.[1] (Needless to say, if you can inspect *all* containers, you have the best sample of all.)

- Each inspection must follow the same, exacting procedure so that each type of contraband has exactly the same likelihood of being detected in each inspection regardless of who is executing it.

---

[1] This is the simple version of the rule. The complicated version of the rule would take into account a technique called stratification—e.g., if I believe that the persona of the shipper is more important than the individual container, a more efficient sampling strategy might randomly select five containers from each ship regardless of the ship's size. While this means that the probability of a particular container to be checked is a lot smaller if it is on a Triple E class ship (which can carry up to 18,000 containers) than if it is on a small vessel carrying just 250 containers, it is still random, and by applying weights (capturing for each container what fraction of the ship's total cargo it represents) we even can undo our stratification when estimating the coefficients of the algorithm.

If you think through how to pull this off with 100,000 customs inspectors dispersed across countless ports, land crossings, and airports, you realize that this is a real managerial challenge. In general, there are two important techniques to find a solution to this kind of problem:

- Before collecting any new data, there is both a real need for and real value in defining what a standardized "best practice" inspection routine would entail. How to do this really would merit a book on its own—but in a nutshell, you would:

    - Collect insights from your best inspectors as to which inspection techniques are most effective (e.g., you may learn that some inspectors have unofficial cheat sheets for each major brand of designer handbag that list which details are most likely to give away fakes).

    - Use their collective wisdom to prioritize among all the ideas you have collected a manageable number (e.g., 10 to 25) of specific work steps that will define the standard routine of your best practice approach going forward (this may contain context-specific items that are only applicable to particular types of shipments, or you may even have completely separate approaches depending on shipment category).

    - Define for each work step how best to execute it in order to avoid human biases (including ego depletion caused by mental fatigue).

- Rather than trying to mobilize your entire army of front-line personnel (which would need to be convinced of the merits of your new approach, trained in its execution, and monitored for compliance), it is often a lot more realistic to train up a small team of inspectors who execute your new approach under close supervision on a carefully selected sample of shipments.

The result will be a truly unbiased dataset that allows you to build an unbiased algorithm to steer investigations in the future. In fact, you have an opportunity to not only predict which container may contain contraband but you could even develop algorithms to predict for each of your, say, 25 best practice work steps the expected success rate for a particular container. This model design would be extremely efficient because going forward, you can suggest to investigators not only which containers to inspect but also which handful of checks to perform (omitting, say, the laborious inspection of the L's, O's, and T's on a leather stamp if the only concern is that the shipment papers might have understated the quantity of bags). However, be aware that this model design will introduce a new bias to the data, and next time around you will have to run a new experiment to once again collect truly unbiased data.

My key point here is that the generation of unbiased data is quintessentially a managerial challenge with some technical guidance by a data scientist on the margin. If your customs inspectors don't follow your glorious new standardized approach because their biases get the better of them ("Why should I waste my time on shipments of Zeta Reticulans—it's *always* the Martians who smuggle stuff," they might grumble), you'll suffer a *déjà-vu* and get as badly biased data as you had before.

Of course, this customs inspection example is by all means a bit of a worst-case scenario. In many decision processes, you will find that it is a lot easier both to obtain unbiased measures of outcomes and to create true random samples. For example, in credit scoring, it is a best practice to have the computer running your algorithm randomly select applications that are approved regardless of what your scoring model says (thus creating a true random sample). The outcome is objectively determined by whether or not the customer pays back the loan—no subjectivity involved.

Such a loan sample is, of course, expensive (you would expect a pretty high number of defaults and as a whole, the sample may destroy money because you lose more than what you earn back through interest), and you therefore will carefully limit its size to the minimum your data scientist needs to improve the algorithm (I usually see my clients run tests of 500 to 2,000 loans at a time). You also can avoid unnecessary losses by still excluding applicants who you are certain to be unworthy of credit, such as applicants who are currently already in default with other banks. In spite of its cost, however, this approach is the gold standard to mining the new gold—unbiased data.

And in order to combat dynamically arising biases, it is critical to not just generate unbiased data in a single, one-off effort, but to embed it as a regular practice into the way you run your business "as usual." Many situations introduce hidden biases. For example, if a bank rejects a certain profile of customers, it does not generate data that would prove the algorithm wrong (i.e., there is no record on the customer showing that she repaid the loan she had applied for), and there may not be external data for proper reject inferencing either (e.g., if *all* banks refuse to lend money to Martians, the credit bureau will lack any evidence of Martians repaying their loans). Only the regular generation of unbiased data on Martians will keep your algorithm honest—and through this also spot business opportunities to give profitable loans to creditworthy Martians.

Summary

In this chapter, you learned how thoughtful efforts to create unbiased data can mine a new type of gold. Key take-aways include:

- When buying a Louis Vuitton bag, check the L's, O's, and T's of the leather stamp in order to make sure it is not fake.

- Readily available data for modeling purposes can be lethally biased, and the only way to obtain unbiased data then is to run a careful process redesign or to conduct a pilot that operates a completely unbiased data collection.

- In order to obtain unbiased data, the data must comprise either all cases (i.e., the entire population) or a true *random* sample, and the assessment of the outcome must follow a uniform, standardized approach across all cases.

- If humans are involved in the generation or collection of the data, executing such an unbiased data collection is a challenging effort because it requires complete compliance by the front-line. It therefore often is best undertaken as a pilot involving only a selected group of front-line staff and a carefully constructed sample.

- Automated processes, by contrast, lend themselves to regularly running trials with completely randomly selected cases to challenge and continuously improve the algorithm.

- And only organizations that embed the regular generation of fresh, unbiased data into their "business as usual" operations can ensure that their mine of data gold never is polluted by biases.

With this, we have concluded our discussion of algorithmic biases from a user perspective. We started with a top-down look at the overall architecture of a decision-problem and a discussion of whether and in which shape algorithms should be applied in a particular decision process given their risk-benefit profile. We then proceeded to review safeguards for the use of algorithms and ways to monitor them. Finally, we discussed how managerial strategies can overcome stubborn biases embedded in algorithms and how to create new, unbiased data to develop unbiased algorithms. In the last, fourth part of the book, we now will turn to the data scientist and review in more detail the specific techniques a data scientist can use to eliminate or reduce algorithmic biases.

What to Do About Algorithmic Bias from a Data Scientist's Perspective

The Data Scientist's Role in Overcoming Algorithmic Bias

As our journey through the world of algorithmic bias has shown, data scientists are facing a formidable challenge, where age-old societal practices and biases, business owners, users, naughty datasets, and the tired brain of the data scientist all might conspire to introduce algorithmic bias. At the same time, data scientists have a lot of powers to contain algorithmic biases through thoughtful modeling choices. In this final part of the book, we will discuss in greater and more technical detail the most important techniques for data scientists to contain algorithmic biases. Lest I am creating unrealistic expectations, let me stress: algorithmic bias is not a single foe but a whole army of adversaries—building a model can feel like a trek through a jungle where you need to fend off everything from mosquitos to poisonous snakes and ferocious tigers. Therefore, sadly, there is no silver bullet, just a survival class and a packing list of recommending essentials such as mosquito repellent and a machete.

© Tobias Baer 2019
T. Baer, *Understand, Manage, and Prevent Algorithmic Bias*,
https://doi.org/10.1007/978-1-4842-4885-0_18

In this chapter, we will revisit the model development process introduced in Chapter 4 and call out specific steps and tools for the bias-conscious data scientist. The next chapter will dive deeper into techniques to assess whether any hidden biases lurk in the data. Considering that machine learning can be both friend and foe, Chapter 20 will discuss when machine learning is a foe and should be replaced by more manual techniques, while Chapter 21 discusses how machine learning can be your friend and help in debiasing algorithms. Chapter 22 revisits the issue of self-improving machine learning models on a more technical level, and Chapter 23 takes a governance perspective and discusses how to embed the techniques discussed here in a large organization where you have more data scientists than you possibly can personally guide and supervise.

Step 1: Model Design

In model design, I recommend in particular three practices to counter biases:

1. Get real-life insights from the front-line! If you are modeling credit defaults of family-owned grocery stores, talk to the credit officers assessing them, the sales people in the branch selling credit to them, the collectors chasing them for unpaid loans, and the sales representative of a famous Italian company selling irresistible, unhealthy confectionery to them. If you are modeling motor insurance claims, talk to the police about car thefts, talk to mechanics in body shops about repair costs, and ask the folks of a German or Swedish automobile club who conduct the so-called *moose test* (defined in the international norm ISO 3888-2) about vehicles toppling over when drivers swerve around hairy beasts jumping on the road.

2. Consider whether the data available to you now might be tainted by a real-life bias, and how best to address it. Maybe you can find external data to correct the bias (e.g., for rejected loan applications in many markets data to conduct so-called reject inference (i.e., figuring out what would have happened if the loan had approved based on the customer's behavior with other banks) can be bought from credit bureaus), or maybe you must push for a dedicated data generation effort (e.g., an experiment involving randomly selected cases) to create unbiased data from scratch.

3. Ensure that the overall business design prevents fatal feedback loops.

I cannot stress enough the need to talk to people "out there" on the front line who deal with real life every day. Their insights are invaluable especially to avoid biases by omitting true predictive factors (while including irrelevant factors that trigger biases) and to derive a definition for the dependent variable (i.e., the outcome you will model, such as what is good versus bad) that preempts biases. The frontline also will be able to tell you about historic events that might have biased data in the past (traumatic events affecting certain time periods and/or locations) and provide you with top-line benchmarks (such as totals for the population you are modeling or key ratios such as average consumption per capita) that you can use to check whether your sample appears complete and consistent with real life.[1]

When talking to the front line, be thoughtful about which questions are more or less likely to elicit biased answers. For example, it is much easier for a credit officer to correctly judge which company attributes in general make a company "more profitable" than to judge what makes it "more likely to default." He will have seen both profitable and unprofitable companies by the dozens (and hence have a solid experience base to observe differentiating attributes) while the difference between a low and a high likelihood of default may be as little as 3 per 100 companies, and hence for many attributes the credit officer is extremely unlikely to have observed enough cases for a statistically valid assessment; instead, his answers are likely to be anchored in a couple of traumatic experiences (e.g., if a US bank had a single German company as customer and this company defaulted in a spectacular collapse that caused a lot of unpleasant questions for the credit officer,[2] his amygdala is extremely likely to cause him to blurt out "Germans!" when asked what marks a risky company).

I also cannot stress enough how critical it is to think long and hard if the readily available data might be biased (e.g., recall our customs inspection example), and if necessary, to push hard for creating new, unbiased data rather

---

[1] As a side note, people who have "skin in the game" (i.e., make decisions based on their hunch about these risks that will affect them personally) often develop correct subconscious impulses (using the brain's "machine learning" capabilities) long before their logical minds can articule where the risks lie and why—for example, in experiments where gamblers got to draw cards from two different stacks where one stack had better odds of winning, their hands "knew" the better stack long before their conscious mind. You therefore might strike gold if you can observe actual *choices* people with skin in the game make and then reverse-engineer what is driving these choices.

[2] The fraudulent German real estate developer Jürgen Schneider comes to mind; one of his "masterpieces" of sorts was a building that didn't have one floor above the other like traditional floors but rather had one single floor that spiraled up—and as a result nobody noticed that the actual floor space of the building built was just a fraction of the stated floor space on which valuations were based. You can imagine that the poor credit officer who signed off on this valuation would never want to touch a piece of German architecture again!

than working with garbage data. To put this into perspective, when I interview data scientists, the single most frequent reason why I reject a candidate is because he or she overlooked that my case study involved biased data (even though I threw in blatant hints). And pushing for an effort to collect new, unbiased data is really hard for three reasons: it very often means that a deadline the business has in mind will be missed, it requires an investment (of both money and management time), and decision-makers have an innate bias to minimize short-term pain (i.e., avoid blowing a deadline or asking their boss for an investment) over long-term pain (i.e., being stuck with a biased algorithm). The best way to overcome this bias is to ask the decision-maker to imagine how in a year's time he has to defend himself because the new algorithm he has commissioned is completely useless and his boss wants to fire him because he has failed to fix the problem.

To illustrate the gravity of the problem, let me recall the fate of an Asian bank that built a new statistical model for consumer loan underwriting. The model was built using the loans the bank had originated—and nobody had alerted the statistician that historically the branch staff had sent away customers whom they had *deemed* to be risky (e.g., one of the high risk groups were policemen—in this market, they believed themselves be above the law and therefore tended not to repay financial debt). Once the new model was in place, the branch staff was asked to run *all* applications through the new system, and to their delight (and probably surprise, too) most of the "risky" customers were approved (e.g., policemen). The result was a huge increase in credit losses—the new model had destroyed millions of dollars. The *only* way for that bank to have avoided this fate would have been to record at least for a sample (e.g., all applications sought in a couple of branches over a two-week period) *every* customer that came "through the door" (with all the information provided in the application) along with what the front line did (send the customer away or proceed with the application).

Finally, the consideration of potential feedback loops reminds us that algorithms are not ends in themselves—we build and use them for a higher-level business goal, and the algorithm a business owner has requested and the way he intends to use it may or may not be suitable for achieving his goals. Here it is our opportunity and duty to be a thought-partner that guides the business owner to designing a better business process and help define the proper role and design for a single or entire suite of algorithms to support this process.

A fourth consideration in model design is the choice of modeling technique in light of potential biases. This topic we will discuss in detail in Chapter 20.

Step 2: Data Engineering

In Chapter 4, we discussed five steps in data engineering, and for each step I would like to call out one practice to prevent algorithmic bias:

- Be paranoid about **sample definition**—for almost every model I build, I find sample definition the single most challenging task because it entails so many traps that cause biases. I in particular worry about (1) *stratification* (i.e., how to navigate size limitations on my sample due to cost and time budgets that threaten to yield an insufficient number of cases for specific profiles), (2) which *time periods* to cover, and (3) the overall *sample size* required to achieve a robust model (considering the number and size of subsegments which I hypothesize to have distinct behaviors, the frequency of certain rare attributes, and the modeling techniques I intend to use). All of these decisions usually require creative thinking to minimize cost (e.g., I may end up creating a modular structure where I have a smaller sample for a singularly expensive data source and a much larger sample for all other data sources) and may require asking for a larger budget (e.g., in order to buy external data on a larger sample).

- **Data collection**: Be extremely vigilant about queries or manual processes that inadvertently could exclude/miss certain profiles—both due to technical issues (e.g., for a certain overseas branch, the off-line upload of certain data may be delayed, which explains why a year-end snapshot pulled from a data archive might exclude cases from that branch) and conceptual problems (e.g., in many contexts, there is a survivorship bias). For example, when you ask banks for a list of all defaulted accounts, you run a very high chance of missing cases. I had banks that purged written off (i.e., unpaid) accounts from their IT system; removed the default flag from the system when they agreed with the customer to restructure the loan (even if at a loss to the bank); or reported the paper file with essential loan information "missing" for defaulted customers because the file had physically been moved to their work-out department. Similarly, SQL queries can easily have tiny conceptual flaws that cause them to miss certain profiles (or misclassify, duplicate, or scramble them). Again, the key to all of this often lies with the front line: quiz senior accountants and back-office staff in the collections department about the gazillion ways a loan can go "bad" and get useful benchmarks (e.g., total credit loss booked for the portfolio in question in last year's financial statement) to gauge whether your dataset looks complete.

- When **splitting the sample** into development, test, and validation samples, choose time periods carefully, paying attention in particular to truncated performance windows for the outcome and truncated data histories among predictors, two formidable sources of conceptual bias.

- **Data quality** is so central to avoiding bias that I have dedicated the entire next chapter to this topic.

- In **data aggregation**, test the conceptual soundness by carefully examining a couple of test cases, especially for exceptions. An example of typical issues are data elements with a 1:n (i.e., one-to-many) relationship—for example, a bank might find multiple loans of a customer in a credit bureau report; a doctor inquiring about the history of cancer in a patient's family might encounter multiple siblings, uncles, and so on; and a customs inspector looking up a shipper's history might find multiple shipments in the past. How do you summarize these? For high-risk markers, often the default choice is a "worst of" criterion (e.g., "any (or ever) 90 days past due"). However, this introduces a subtle bias against proliferations: if you are blessed with an extremely large family where for generations your family prided itself in having double-digit numbers of children, everything else being equal you are a lot more likely to find a single cancerous case in the family history than if your ancestors were avid advocates of a single child policy! Hence in an algorithm tasked with assessing the likelihood of cancer, the variable "worst cancer status in the family" could create a bias that overestimates the probability of an individual to develop cancer if the individual comes from a very large family and underestimates if the family is large. At the same time, a ratio (% of family members with cancer) is similarly misleading—if you recall earlier discussions of significance, you will remember that 0 cases out of 5 is a lot less meaningful than 0 out of 500. In this case, the best approach would be to create two aggregate variables—number of family members with cancer and the number without—and to create more complex derived features from these values during feature development (which we will tackle in the next step).

Taken together, these five substeps within data engineering can eliminate many biases in the data—but some biases still can survive at this stage and need to be caught in model assembly.

Step 3: Model Assembly

Within model assembly, we distinguished seven steps in Chapter 4. In each step, a data scientist can take specific measures to keep biases in check. Many of the biases—and the appropriate ways to combat them—are extremely context-dependent, however; the best I therefore can do is to call out in each step what kind of biases to target on a conceptual level and illustrate with examples what this might look like in real life. I urge you to consider in each step how best to obtain as much context information as possible, and whom to leverage as thought-partners to brainstorm potential biases and ways to eliminate them; as the following discussion hopefully makes clear, it is impossible to reign in algorithmic bias without plenty of such discussions!

- **Exclusion of records**: What kind of records in the sample would be misleading to the statistical development of the algorithm? Credit applications are a good illustration because there can be as many as four types of misleading cases present (if not more):

  - **Immaterial defaults** (i.e., tiny amounts) typically happen due to other reasons than credit problems, as discussed above.

  - **Identity fraud** cases are misleading because the applicant's profile was irrelevant and the loss occurred because the bank's identity verification failed (and hence a criminal posing as someone else with a good risk profile could get the loan in the other person's name).

- **Inactive accounts** (e.g., closed or frozen credit cards, or car loans that were approved but never disbursed because the customer cancelled the car purchase) do not entail an actual payment obligation of the customer (they are like ghosts in the bank's systems without any real money ever having been handed to the customer) and it is therefore impossible for the customer to default.

- **Special segments** are those where normal decision processes do not apply (e.g., development loans with a government guarantee—here the bank's risk assessment is irrelevant and the only question is whether the application meets the government's requirements).

- **Feature development**: Here the data scientist should be worried in particular about two issues—omissions and conceptual misconstructions.

  - Do you *miss* important predictors of outcomes (especially those suggested by the front-line people)? If so, what features do you have that might be correlated with the missing features? Could they lead to biases in those cases where the correlated features are poor proxies for the missing features? If so, how could you possibly collect the missing features, even just for a small sample? If this is absolutely impossible, is it responsible to develop the biased model? What kind of overlay over outputs would help to prevent this bias from causing damage?

  - And are derived features *conceptually sound* in *all* situations? For example, by evaluating a ratio for the minimum and maximum values of the two input variables present in your sample, you might stumble over negative values, which can screw up a ratio that you have calculated. In fact, ratios are a great example for seemingly simple aggregations that have lots of traps. For illustration, let's consider a standard ratio used for assessing the health of companies: the gearing or debt-to-equity ratio, commonly defined as total debt divided by equity. Normally a smaller value is safer ($10/$100 is 0.10 and represents a lower level of indebtedness than $50/100 (=0.50); by contrast, $10 million/$100 million is also 0.10 because this is simply a larger company but relatively speaking it has the same amount of debt per unit of equity as the smaller company—ratios are used precisely because they measure structure but ignore scale). However, some struggling companies have negative equity—the larger their *debt*, the bigger their problems but the smaller the ratio of debt over equity (which is approaching minus infinity for an infinitely large debt). Furthermore, many credit analysts find that a better metric of debt is *net debt*, which deducts from total debt all cash and cash-like assets (e.g., treasury bonds). If a company has more cash than debt, net debt can become a negative value, too. Dividing negative net debt by positive equity (an extremely healthy company) also yields a negative value for the ratio—hence a ratio of −0.5 could denote either a very bad company (lots of debt and negative equity) or a very safe company. And if the latter type dominates the sample, the model easily could have a positive bias towards companies with a negative net-debt-to-equity ratio and enthusiastically approve some very risky companies, causing huge losses for the bank. To solve this, you need to treat all the different combinations of positive, negative, or zero values for denominator and numerator separately. Advanced machine learning techniques are usually reasonably good with this but if such cases are rare, nonsensical treatments still can slip through. Alternatively, the data scientist needs to create a complex new feature that carefully maps the various value ranges of denominator and numerator into a continuous risk index (that puts negative debt and positive equity into the safest region, normal value ranges into the middle, and companies with positive debt but negative equity into the riskiest region, etc.).

- **Short-listing**: Once you have a draft short-list of features, you should go back to the front line to get feedback on whether any of the features could cause a bias (e.g., you might believe that having a luxury backpack brand is a great predictive feature for positive outcomes but a Martian might tell you that this would discriminate against Martians because most Martians believe that they will be singled out by Zeta Reticulan police for frequent searches while passing through train and subway stations if they carry a backpack and therefore avoid them). Often there is an opportunity to swap problematic features for less problematic ones that are highly correlated. The Principal Component Analysis is a great tool to facilitate the search for alternatives, and this work step is the ideal time for this discussion because a shorter list of features allows for more in-depth discussions (as opposed to the initial data collection, which may entail thousands of data fields) but you still have all degrees of freedom to replace variables.

- **Model estimation**: Here you should carefully check hyper parameters to avoid overfitting—the source of harmful model artifacts.

- **Model tuning**: This is the time to run the full gamut of analyses we discussed in Chapter 15 to detect biases in the initial model. I also always urge my data scientists to *visually inspect* all features in the final model by plotting the feature against (moving or category averages of) log-odds ratios (for binary outcomes) or the continuous predicted output (at least the top 10-20 features if the number is large)—it is a great tool to identify problems in the feature development. In addition, you can *simulate* model predictions for a couple of representative cases (real or artificially created) against which you suspect the algorithm to be biased. As an example, I really liked it when a board member of a Taiwanese bank I served pushed me on whether the new model might unfairly discriminate against entrepreneurs. He laid out the example of a beauty surgeon who had substantial professional experience in the US but was opening a new clinic in Taiwan and approached the bank for funding. We created a hypothetical application for a beauty surgeon and were pleased to see that the model approved it, albeit with an elevated risk estimate that we felt adequately reflected the risks involved with a new business.

- **Calibration**: Here your attention should turn to two issues: stability biases and subsegments with poor predictive power. To prevent *stability biases*, debate how the future could be systematically different from the sampling period, and how the central tendency (e.g., the average default rate of a credit portfolio) could develop in the future. And in order to prevent algorithmic biases against *subsegments where the model has poor predictive power*, decide in which cases to relabel model outputs as "don't know" rather than providing estimates that are a shot in the dark.

- **Model documentation**: This task is strongly disliked by most data scientists; it is even skipped at times if there is no strong governance in place. Part of the challenge might be that writing dozens of pages of text is a different skill-set than modeling and therefore is often neither easy nor pleasant for data scientists. More fundamentally, however, it is often perceived as a low "value add" and exclusively for the benefit of others, not least formal validators (who tend to be disliked, too). This is a great shame because if it is suitably designed, model documentation can be a great tool for data scientists to build better models and avoid biases! In order for model documentation to achieve this objective, it must meet two requirements: it must have a format that triggers helpful reflection, and it must be written "on the go" while insights arising through the writing of the documentation still can trigger an adjustment of the most recent work step (reflections by the data scientist on sampling are rather useless once the model is built). A great way to achieve this is to structure the model documentation in a Q&A format; it basically becomes a recorded discussion between the data scientist and a fictitious friend. In Table 18-1, I provide an example of a table of contents of such a model documentation.

Table 18-1. Example Table of Contents for a Q&A-Style Model Documentation

Section	Question
I.	Model design
1.1	What business problem does the model address?
1.2.1	Which sources of expertise/experience were tapped into to identify potential drivers of outcomes?
1.2.2	A priori (i.e., before running any analyses on this), what do we expect to be the main drivers of outcomes?
1.3.1	In what situations will the model be applied?
1.3.2	When will the model *not* be applied (exceptions)?
1.3.3	What will be the process for applying the model? How might this process potentially affect the functioning of the model?
1.4.1	What outcome does the model predict?
1.4.2	What biases are (or could be) present in the historical data used for the model development?
1.4.3	What was done to obtain modeling data without such bias?
1.4.4	What remaining biases are expected to still be present in the data? What usage restrictions for the algorithm does this imply?
1.5.	What is the overall structure of the model?
2.	Data engineering
2.1.1	What objectives, issues, and constraints were considered in the sample definition?
2.1.2	What sample was used for model development (e.g., describe time periods covered, sample size, and any stratification or random-sampling)?
2.2	What issues and constraints were anticipated and/or encountered in data collection? How were they addressed?
2.3	How was the sample split into development, testing, and validation samples? What considerations drove this split?
2.4.1	A priori, what data quality issues were anticipated?
2.4.2	What was done to detect data quality issues?
2.4.3	What data quality issues were identified, and how were they addressed?
2.5.1	What a priori insights or hypotheses about the modeling problem have impacted the data aggregation strategy?
2.5.2	How was data aggregated?
2.5.3	What biases might have been introduced through the way the data was aggregated (regardless of whether later such biases were found or not)?

(continued)

Table 18-1. (*continued*)

Section	Question
3.	Model assembly
3.1	Which records should be excluded on a conceptual level and why? How was each exclusion executed (especially in terms of identifying the relevant records)?
3.2.1	Which features you a priori expected to drive outcomes were missing in the data? Which proxies did you use?
3.2.2	What design principles informed the definition of features (e.g., in terms of normalization)?
3.2.3	Which features have been coded? Which potential features were omitted, and why?
3.3.1	How were features short-listed?
3.3.2	Which features were short-listed for model estimation?
3.4.1	Which modeling technique(s) did you consider, and which one(s) did you choose?
3.4.2	How did you set the hyperparameters for the modeling technique(s) chosen?
3.5.1	What results did the testing of the initial model yield in terms of stability and biases?
3.5.2	How did the model evolve during tuning?
3.6.1	What consideration informed your calibration approach?
3.6.2	What assumptions and approach did you choose for calibration?
3.6.3	For which subsegments do model outputs require some form of qualification (e.g., replacing estimates with "don't know")?
4.	Model implementation
4.1	What issues could arise in model implementation that cause model outputs to be biased?
4.2	What prescriptions would you like to make for model implementation to ensure that the model functions properly and is only applied to situations where it is safe and competent to use?
4.3	What testing program do you recommend to ensure that the model is correctly implemented?
4.4	During ongoing use of the model, what biases could arise over time?
4.5	What ongoing monitoring regime do you recommend to identify such biases and other performance issues?
4.6	What measures must be pursued (e.g., regular randomized experiments) to ensure that there will be unbiased data for challenging and refining the model in the future?

Another way to look at the model documentation template given in Table 18-1 is to consider it a checklist against algorithmic bias. If you are keen to incorporate all the techniques you have learned in this book in your daily work, make it a habit to go through this questionnaire along your way of developing a model—even if you don't use it for documenting the model.

Step 4: Model Validation

A time-tested practice to contain biases in important decisions made by human judgment is to designate someone with an explicit mandate to challenge the decision-maker, a role sometimes called the "devil's advocate". Even medieval kings employed a dedicated staff member to challenge them and provide new information that they may have missed in their deliberations: the joker. In modern model governance, the role of challenger to ensure the soundness of an algorithm often is formalized through a function called model validation. Unfortunately, the joyful nature of medieval jokers somewhere has gone lost, to the great detriment of effective validation.

Based on everything we have learned about even data scientists being biased (I'm no exception—I have a bias to see a bias everywhere, even where maybe just a random error has been made!) and the manifold ways naughty little biases can creep into an algorithm, there should be no doubt about the value and importance of some independent challenge of an algorithm—and indeed I have seen numerous cases where model validation uncovered critical shortcomings.

Sadly, in many organizations—especially financial institutions where model validation is often mandated by law and pursued with particular zeal—model validation has become a rather double-edged sword, if not outright dysfunctional. Overzealous validation functions can make it all but impossible to get a new algorithm approved.

For example, I once worked with a bank where for years not a single new model passed validation. One of the key hurdles was that when the validation team rebuilt a model from scratch (going back all the way to extracting the data from source systems—per se a very thorough and valuable exercise given the dangers for biases in data engineering we have discussed), it required model coefficients to be *perfectly* identical (maybe comparing coefficients for 8 or 10 digits after the decimal point). As a result, even the slightest quantum of dirt in the data—maybe a single number among thousands of cases where a minor typo was made—would fail a model.

If model validation is overly strict, it can cause real damage. For one, an inferior model may not be replaced with a new, superior one because of a concern blown out of proportion. In addition, I have observed that if data scientists are overly afraid of their models failing validation, their behavior changes in unhelpful ways. They can become reluctant to try out innovative ways because they believe the easiest way to get a model validated is by noting that the approach is identical to another, previously validated model; they also may try to conceal weaknesses in the data or elsewhere in the model.

Part of the reason is the unhelpful construct of the validator's role—never has a bank credited its growth or profitability to their model validators signing off on great new models, but more than once have validators been blamed if a model malfunctioned or a regulator raised a concern with a model. The result is a culture where validators have an asymmetrical interest in pointing out even the smallest issues and withholding sign-off even if only the faintest concern arises.

There are a couple of core principles for designing a constructive validation function:

- Rather than a binary outcome (approving or rejecting a model), the outcome of model validation should be a quantification of model risk (e.g., using a traffic light system as simple as green/amber/red, or maybe 4-5 levels).

- A rating other than green and red (i.e., whenever a model has some non-lethal issues or limitations) should always come with a formulation of reasonable usage restrictions (which may entail monitoring particular metrics and a requirement to suspend a model if certain thresholds are crossed, or a limit to the financial downside, such as a maximum loan amount that can be approved by a shaky credit scorecard).

- The evaluation system for validators should include multiple criteria that strike a healthy balance between risk identification, risk prevention, and risk taking. For example a balanced scorecard (that tangibly articulates for each dimension something like five different performance levels) could assess the validator's ingenuity in identifying limitations of a model, the degree to which imposed model restrictions appear commensurate with the model risk identified, and the validator's effectiveness in enabling innovation without incurring undue risks for the institution.

In practice, model validation can be as simple as pointing out a segment for which model results are evidently biased or as elaborate as building a challenger model (i.e., an entirely new algorithm built from scratch, maybe using different features and a different modeling technique) to assess whether the data at hand would also support entirely different conclusions on what kind of outcomes to expect for certain classes of profiles. And humor is always an asset because it can provide cover to even the sharpest intellect!

Step 5: Model Implementation

In model implementation, I want to call out two aspects:

- **Implementation of the predictive features** is an often underestimated challenge. Very often the many data engineering steps required to pull, clean, aggregate, and transform the input data of an algorithm are built up iteratively (e.g., because only during model tuning the 25th data quality issue is identified, maybe requiring the correction of a database query at the very beginning of data extraction). As a result, it can be exceedingly difficult for a data scientist to actually compile a complete set of instructions when the model is done. Furthermore, languages and tools used in production systems (i.e., the IT environment where an algorithm will "live" in real life) are often different from what a data scientist uses during development. I therefore encourage every data scientist to implement (purely for testing purposes) a first version of an algorithm in the production system (i.e., create a "pipeline," in machine learning lingo) as early as possible, and to think through at the very beginning of the model design what tools and techniques should be used in the end to translate code developed during data engineering and feature development (e.g., scripts written in R, Python, or SAS) into a production system that meets the requirement of real-life usage (e.g., response speed or decentral deployment—maybe the algorithm even needs to be deployed on mobile devices). And it goes without saying that post implementation, rigorous testing is required to ensure that no hidden biases have crept in nevertheless.

- **Ongoing generation of fresh, unbiased data** to challenge and improve the algorithm is most likely to happen if it is automated—for example, if a decision-system for loan approval will automatically select a random sample of 100 loans per month that are approved regardless of credit score (maybe after running a couple of common-sense filters). Rather than leaving this to wishful thinking enshrined in the closing paragraph of the model documentation, data scientists should consider it as part of their job in model implementation to make this data generation happen. Think of a new airplane—when it comes fresh out of the factory, it is already equipped with a "black box" that records data to enable the development of even better technology in case it crashes.

Achieving integrity in model implementation and data generation for future model versions are the final bulwark against algorithmic biases.

Summary

In this chapter, you revisited the end-to-end model development process and saw that at every single step, data scientists have specific techniques at hand to keep algorithmic biases at bay. The flip side of this is that at every turn, a bias can slip in if a data scientist lets down her guard. The following are the key strategies we identified against algorithmic biases:

- Invest domain knowledge and real-life, front-line insights into whatever you want to model to inform your hypotheses where biases might lurk and what should drive the outcomes you are modeling.

- Never give into the temptation of using biased data for model development if you know that the right thing to do would be to generate new, unbiased data from scratch—even if that means missing a deadline or settling for a small sample and therefore a much simpler modeling technique than what you looked forward to.

- Understand the end-to-end business process in which your algorithm will be deployed and how that process could harm your algorithm.

- Be paranoid about constructing a good sample, considering in particular a need for stratification, which time periods to cover, and what minimum sample size to push for.

- Be similarly paranoid about the conceptual integrity of your data collection approach—use benchmarks (e.g., expected sums or averages) to double-check that no cases have inadvertently been omitted.

- Do not lose the conceptual soundness of your data through inappropriate aggregation steps.

- Exclude misleading records that are likely to bias the model.

- Be vigilant that individual features will not bias the algorithm because they are either poor proxies for missing features or conceptually flawed—especially after short-listing a manageable number of features, you also should enlist the help of front-line practitioners for this.

- Set hyperparameters carefully to avoid overfitting, and thoroughly test initial model outputs for biases that need to be addressed in model tuning.

- Consider adjustments necessary in model calibration, such as changing the central tendency to overcome a stability bias or overwriting model outputs for segments with low predictive power with "don't know."

- Use model documentation systematically as an enforcement mechanism to "check off" all best practice debiasing steps.

- Create a value-adding model validation function that is focused on the *degree* to which model limitations impair outputs and the formulation of use restrictions that *compensate* for an algorithm's biases, and is *motivated* to strike the right balance between fostering innovation and protecting the organization.

- In implementation, ensure that not only the coding of the algorithm and its input features is flawless but also that a mechanism is in place to generate unbiased data for future model redevelopment efforts.

If you feel like these are more points than you can remember, your hunch is correct—humans barely can remember more than three to five items at a time. So to make this very simple, here's the one-sentence summary of this chapter: Thou shall print the (Q&A style) best practice model documentation table of contents in Table 18-1, frame it, put it on your desk, and follow its flow of thought from now on for every model you build!

And in the next chapter, we will double-click on data engineering and discuss even deeper how to detect biases in "the new gold" (for the unpretentious, that's *data*).

An X-Ray Exam of Your Data

In this chapter, we will dive into the question of how you can detect seeds for algorithmic biases in your data. As must have become clear from the previous chapters, we are chasing many different foes; therefore, we need to scan our data for many different types of potential issues, just as an annual health check might include a dozen procedures to check blood, urine, and various organs. With the recommendations in this chapter, my goal is to give you "a thousand eyes and a thousand ears" in six fairly easy and efficient steps. These analyses will create a set of maps where each map attempts to shade in bright red specific areas of concern, just like how an X-ray exam would reveal broken bones, ruptured organs, and swallowed cutlery. This will enable you to review all significant irregularities and (considering your context knowledge and what you have learned in this book, especially the previous chapter) decide whether there is reason for concern, and if so, what best to do to avoid an algorithmic bias.

While I tried to make the six steps as systematic as possible, I need to stress that this is an iterative process—especially where you have a large number of independent variables, you may do a rather superficial run of some analyses first, and if a variable is picked as a contender for the final model, you may want to redo some specific analyses in greater detail (e.g., digging deeper into outliers or certain missing values), especially if your expertise-driven "bias radar" gives you a hunch that something might be amiss.

© Tobias Baer 2019
T. Baer, *Understand, Manage, and Prevent Algorithmic Bias*,
https://doi.org/10.1007/978-1-4842-4885-0_19

In fact, it can make sense to run the entire X-ray twice—the first time when you have received the raw data and the second time when you have created all your derived features (e.g., aggregations of much more granular data or complex transformations) as your feature generation work may have introduced new seeds for biases.

Through all of this, it is important to keep a critical mindset. The best data X-ray routine is useless if you fall prey to the confirmation bias and explain away warning signs by assuming that it is a spurious effect ("oh, my data is too sparse/old/recent/noisy") rather than trying to find further evidence for a problem if the initial X-ray raises a warning flag.

Step 1: Sample-Level Analysis

With an efficient analysis approach in mind, we start with five sample-level checks that give you a "top-down" perspective on the presence of a wide range of data issues that would cause algorithmic biases:

1.   Do the **count of observations** (e.g., number of loans or customers) and the **sum of key attributes** (e.g., loan amount) match external benchmarks (e.g., portfolio statistics from the bank's management information system)? If not, your sample looks incomplete.

2.   What percent of records have **missing values** in critical fields (e.g., good/bad indicator or identifier)? To state the obvious, the definition of *critical* is that no missing values can be accepted. For other variables, we will examine missing values in Step 3 in a very efficient way. However, you may already want to have a glimpse at the "key" features you expect to be most predictive to check if the missing value rate is within the range you expected.

3.   What percent of records have **zero values** in critical fields (e.g., loan amount)? For some variables, zero is just as bad as missing; for others, it can be a valid value (e.g., the account balance of a loan that was approved but never got disbursed is zero) but indicates a need for special processing (e.g., here the exclusion of the record). In this step, also look at the count of associated records in separate tables with a 1-to-many relationship—for example, if a loan can have multiple (so-called "joint") borrowers and you received a separate data table with one record for each borrower, then an orphaned loan with zero borrowers attached to it would be a problem, too.

4. Are there **duplicates**? It might sound like a rookie mistake but little bugs in queries do return duplicates every once in a while...

5. Is the **average outcome** (i.e., average of the dependent variable/label, such as the sample bad rate for credit scoring data sets) realistic?

These analyses indicate flaws in the data that are likely to cause a bias if they are not addressed; remediation often requires identifying a flaw in the data collection process and doing it again, to the great delight of everyone involved.

Step 2: Data Leakage

Data leakage refers to the pollution of predictive variables with hindsight information (i.e., information that only becomes available after the outcome to be predicted has become known, too). A blatant case of data leakage is the collection of current data to predict historic events (e.g., if I retrieve the current credit bureau report of a customer to predict defaults in the past, such past defaults obviously would be listed in the credit bureau). Most cases of data leakage, however, are a lot more subtle. For example, such hindsight bias often slips through if an attribute is deemed static but does get updated in the precise event that the algorithm tries to predict. For example, loan accounts typically are assigned to a branch; if the central workout department is treated as a separate branch, the default of a loan might systematically trigger the reassignment of the loan to the workout department, and hence the branch identifier now becomes a perfect indicator of defaults. Precisely because such attributes are *deemed* constant, they often do not get archived properly.

Hindsight bias also can hide in missing values. For example, I remember a bank where defaulted loans were purged from the main account system, and missing values for attributes only stored in that system thus became near-perfect predictors of default as well. Many modeling techniques (especially machine learning) treat "missing" as a valid value of an independent variable, and it therefore is not unusual for an otherwise irrelevant variable to obtain a lot of weight in a model simply because whether or not it is missing is a highly predictive indicator.

I propose two analyses to catch data leakage:

1. **Univariate predictive power** of each variable; unusually high predictive power typically is an almost certain indicator of leakage.

2. A test if data for each variable is **missing at random**.

Measuring Univariate Predictive Power of a Variable

For assessing the predictive power of an individual variable, I use two or three metrics because I am trying to solve an asymmetric problem; I am concerned that I might miss an issue if I use only one technique that turns out to be biased against a certain type of distribution of a predictive variable (thus failing to recognize its high predictive power) while I am not very concerned about false positives (i.e., incorrectly flagging a variable that turns out not to suffer from leakage—if my follow-up analysis suggests that the variable is OK, I still can use it).

For continuous dependent variables (outcomes), I usually recommend calculating both **Pearson's product moment correlation** coefficient and **Spearman's rank correlation** coefficient. *Pearson's product moment* obviously is the standard correlation metric but it can be biased by outliers and skewed data (i.e., if the distribution of the data doesn't follow a normal distribution). *Spearman's rank correlation* is insensitive to these issues and also works with ordinal but non-continuous values (e.g., alphanumeric grades or university ranks) but it can make a fuzz about clusters of observations with the same or very similar values—within such clusters, the sequence position of individual observations is sometimes not random (e.g., in a data table, you still might find remnants of alphabetic sorting of subjects), and if data processing left some variables ordered in the opposite direction (e.g., Z to A instead of A to Z), biases in the measurement of Spearman's correlation can arise. The assessment of what correlation levels are "suspiciously high" is heavily contextual and should reflect the data scientist's experience with comparable modeling efforts.

For binary outcomes, I typically consider both **Gini** and **Information Value (IV)**.

We discussed *Gini* in Chapter 15; for univariate Gini, we treat an individual variable like a score. This points at two important limitations: Gini only works for continuous and ordinal variables (but not for categorical ones), and it assumes that every variable has a monotonic relationship with outcomes—if the relationship between a highly predictive variable and outcomes follows a perfect U-shape (e.g., credit risk and age: very young and very old people usually are risky, while people in their "best years" have the lowest risk, as they tend to be mature, healthy, and highly employable), Gini can be zero!

Information Value is popular because it also can process categorical values and non-monotonic relationships and because convenient rules of thumbs exist for interpretation; an IV between 0.3 and 0.5 is considered strong, while weak predictors only achieve an IV between 0.02 and 0.1. Values above 0.5 typically are considered suspicious; weaker cases of leakage can only be detected by comparing the IV of a variable at hand with the typical IV for such variables (e.g., while you would expect an IV of 0.3 for a credit score, you would be suspicious if the CEO's zodiac sign achieved that level of predictive power).[1] Another benefit of IV is that its calculation includes the calculation of a value called *Weight of Evidence*, and we will use this value also in a subsequent analysis.

The complication of IV is that it requires bucketing. For example, if you have a variable called "income," you first need to define a limited number of income bands. There is a computational requirement that each bucket contains at least one observation of each kind of outcome (e.g., good and bad). And in order to efficiently assess hundreds or thousands of variables, you typically need something fully automated—as a result, IV often is calculated based on just four to five buckets and is therefore biased against indicators that apply to only a small fraction of the sample.

Assessing If Values Are Missing at Random

For continuous outcomes, I suggest testing whether outcomes are significantly different between observations where a given variable is missing and those where it is not missing (see Chapter 15 and the discussion of the t-test).

For binary outcomes, several equivalent approaches exist to detect whether a missing value conveys hindsight information, in particular:

- Gini or IV of a dummy that is 1 if a variable is missing and 0 otherwise.

- Difference in IV of a variable if IV is calculated once by treating "missing" as a separate bucket and once by excluding those observations where the value of this variable is missing (i.e., measuring the IV of the non-missing cases only).

In all of these metrics, hindsight bias shows up as significant predictive power of the "missing" label.

If missing values convey hindsight information, you need to decide whether to simply drop the variable or whether to fix the problem through appropriate missing value imputation (e.g., a conditional random imputation can fix some situations). For this decision, I recommend looking at the variable's univariate predictive power *excluding* missing values. If that is negligible, the variable has nothing to offer besides hindsight information and should be dropped; if that is still material and appears worth the data scientist's effort, a treatment can be attempted.

---

[1] Of course, if you are an astrologist, you also might conclude that an IV of 0.3 is too low and indicative of some zodiac signs incorrectly coded in your data...

Step 3: Two Analyses to Understand the Structure of the Data

As a data scientist, you obviously want to avoid not seeing the forest for the trees—the same goes for detecting biases. I therefore strongly recommend two analyses to understand the structure of your data in a simple, map-like, top-down view:

- Find **blocks of variables with similar missing value structures** to inform model design (e.g., use a modular model structure if credit bureau data (which might account for hundreds of variables) is unavailable for a big chunk of customers, or consider manual data collection if important variables are *systematically* (i.e., in a non-random fashion) not digitally available for a large number of cases);

- Enable laser-sharp focus of subsequent steps by eliminating redundant variables through a **Principal Component Analysis (PCA)**.

To detect *missing value structures*, you create for each variable a mirror variable that is a binary indicator set to 1 if for a given record, the variable is missing, and 0 otherwise. For example, if for 40% of customers credit bureau is missing and your credit bureau record comprises 140 variables, you would have to create 140 indicator variables that all show 1 for this 40% of the sample and 0 otherwise (the point being that at this point, you don't know this yet—a little script will code this robotically to help you discover this information).

By inquiring the correlation structure of these indicator variables—typically a simple table with bivariate correlation coefficients is both sufficient and most practical (more complex techniques that would also capture multicollinearity might not work because they cannot handle perfectly correlated variables) —you can identify blocks of variables where their missing indicators are highly correlated, which means that they tend to be missing at the same time (and hence are likely to be caused by the same root cause).

The *Principal Component Analysis* was mentioned in Chapter 15 in passing. In my experience, the PCA is one of the most powerful tools in developing an algorithm. Especially with machine learning, there is a widespread belief that adding ever more variables to an algorithm can only improve the performance. The hidden downside is that if you have thousands of variables, you neither have the time nor the mental capacity (before mental fatigue sets in) to examine each variable thoroughly for any hint of bias. The PCA can boil down thousands of variables to a few dozen on which you then can "double-click" to do more detailed analyses (e.g., to detect anomalies), including visualizations of the data and actually talking to front-line people to obtain explanations for any potentially biasing phenomena you observe.

The calculation of IV in the previous step is a critical prerequisite for my recommended approach of using the PCA for two reasons. First of all, in order to be able to run a PCA, you first need to convert categorical variables into numerical ones and impute all missing values; the IV calculation does this conveniently by offering you a Weight of Evidence (WOE) for each observation (in the form of the WOE of the bucket in which a given observation sits—so this is obviously often crude if you have just a handful of buckets). Second, the IV will drive the prioritization of variables within each principal component.[2]

In order to answer the question of which variables are most important to look at (and consider in a model) with a PCA, I suggest you go through the following steps:

- Run a Principal Component Analysis with the cut-off criterion set at Eigenvalue = 2 (a standard rule-of-thumb).

- Perform a Varimax-rotation in order to align the maximum number of variables with the first principal component.

- For each row (each raw variable is a row), calculate the maximum *absolute* loading factor and assign variables to the corresponding principal component.

  - If maximum absolute loading is ≥ 0.5, label the variable as belonging to the principal component with this loading.

  - If maximum absolute loading is < 0.5, label the variable as belonging to "other."

- Sort variables in a two-level hierarchy: first by the number of the principal component to which the variable belongs ("other" comes last) and then by the absolute factor loading; this means you first have all the variables associated with principal component #1 (which has the highest Eigenvalue), then the variables associated with the second principal component, etc.; within each block, you have sorted variables from most to least representative of the principal component (a factor loading of both

---

[2] If the dependent variable is continuous, you do not use IV. In this case, you can create an equivalent imputation table by calculating median outcomes for non-numerical categories (including missing). For prioritization, I use the larger of Pearson's and Spearman's correlation.

+1.00 and -1.00 suggests that the variable is a perfect representation of the principal component; the sign of the factor loading is irrelevant for our analysis).

- For each principal component, pick the variable with the highest IV (your choice also may reflect fill rate and practical considerations such as ease of implementation—often the second or third most representative variable is better than the most representative); amongst equal IV, give preference to the variable with the highest factor loading.

- If you have picked a variable with a factor loading of 0.5~0.8, consider picking another one or two variables for the same principal component if these also have high IVs and based on business judgment you believe that these variables might contain some unique information you have not captured yet in another picked variable.

- Finally, review all variables in the "other" bucket; these variables are somewhat approximated by the principal components but if you see variables that have a high IV (> 0.3), you may short-list them as well, especially if business judgment suggests that the variable would be expected to be in the model and is not well represented by the other variables already short-listed.

It should be clear that this procedure is far from deterministic or perfect; it involves a lot of judgment whether or not to pick a variable. It is, in fact, a blunt, pragmatic approach to establish that there are lots of firs, beeches, and giant sequoias in your forest so that you can check the firs for the presence of any tussock moth caterpillars (a bug that can quickly devastate large areas of firs) first before getting too hung up on the sole peach tree in your sample that seems to have a tiny amount of frass (which could indicate the presence of a Crown borer, a peach-specific pest).

I personally like to build a model with only a few dozen variables prioritized by the PCA at least as an initial benchmark as I then also can really double-down on feature generation where I have additional opportunity to detect hidden biases, and there are only that many variables about which you really can have in-depth conversations with front-line staff and other experts. I found that models built that way often are just 1-2 Gini points below the performance of the most sophisticated machine learning model that uses all of the original raw variables, and that thanks to their superior robustness they often start to outperform those sophisticated machine learning models after a short period of time (as small structural changes start to bring to the fore some of the undetected overfitting and other biases still hidden in the black-box machine learning model).

However, once you have completed your data X-ray, you are at leisure to include as many variables in your further model development as you like. The PCA is a tool very much like a real X-ray (that reduces the three-dimensional body to a black-and-white picture of bones) and not more. If in doubt whether to include a variable in the further modelling work, you probably want to err on the side of inclusion.

Step 4: Two Analyses for Anomaly Detection

In Chapter 15 I mentioned the favorite quip of a good friend of mine: "What's normal?" In spite of the relativity of this question, it is important to make an honest attempt at identifying anomalies in the data—outliers, unexpected concentrations of values, and other tracks of the bias devil.

In the following section, I propose one manual approach for anomaly detection and one using machine learning.

Manual Anomaly Detection

One of the best anomaly detection devices in the world sits right there on your shoulders: your brain. Since ancient times, animals have spotted dangers by scanning their environment for anything deviating from what is usual, and the human brain is no different. Therefore I will show the data to the brain in two steps: here I show numerical outputs, while in Step 6 I show graphs (thus presumably engaging both sides of the brain). However, because of mental fatigue, focus is critical—this is why in the previous step we conducted a PCA in order to shortlist a manageable number of variables before proceeding with this step. Table 19-1 shows the report I recommend for categorical variables (including numeric codes such as industry codes) while Table 19-2 shows the output I suggest to generate for numerical variables (which can be continuous or at least ordinal, such as rank).

Table 19-1. Output Template for Categorical Variables

<Variable name>	<Most frequent value>	<% of records>
Missing: <% of records>	<Second most frequent value>	<% of records>
Number of categories: < # >	<Third most frequent value>	<% of records>
	<Fourth most frequent value>	<% of records>
	<Fifth most frequent value>	<% of records>

Because categorical variables typically contain text, I find a "report" format best especially if some labels are quite long. Here each variable takes up five lines. A blank line between each block and nice formatting really helps the brain! Look at it, mark (e.g., with a circle or exclamation mark) or write down any spontaneous reactions (e.g., surprises, next steps for what to investigate, as well as ideas for feature generation), and move on. If you feel you are getting tired and move on too fast, stop for a break (getting some caffeine and looking at some nature such as a tree in front of the window really helps!), and redo the previous couple of variables that may have gotten short-changed.

Table 19-2. Output Template for Numerical Variables

Variable	Missing	Min	Max	Mean	Median	Mode	Percentiles 1st 2nd 5th 10th 25th 75th 90th 95th 98th 99th
<Var #1>	<%>						
<Var #2>	<%>						
<Var #3>	<%>						
...							

Table 19-2 is a big table with one row per variable. It is possible to go through it line by line but again mental fatigue is your enemy. A better strategy is to break this down in five runs:

- First, sort by **missing value** (percent of observations) from largest to smallest—at the top of your list you thus will see all variables with a high share of missing value, which either might suggest a data collection problem or raise questions for the model design.

- Next, sort by **minimum value** from smallest to largest—you will see negative values first, and every variable where a **negative** or **zero** value doesn't make sense or at least raises questions gets transferred into the "to be investigated" bin.

- Third, sort by **maximum value** from largest to smallest—again a couple of variables may reveal themselves as questionable.

- Then, sort by **mode** and see whether for any variable the mode raises questions—I sort because often a particular problem is associated with a particular value.

  - Zero values are sometimes a kind of missing value (and sorting will put all zero value cases together).

- Large numbers with all digits set to 9 (e.g., salary = 9999999) often were used as missing value indicators in ancient IT systems and still linger in some databases today.

- Round values of continuous variables (e.g., salary = 1000) often are a tell-tale sign of a human being entering a guess instead of the true number.

- Finally, flip through the remaining variables that have not been flagged already as non-normal by the previous steps. Minimum, maximum, mean, median, mode, and the percentiles all give you a sense for the distribution—does it make sense? Do the lower percentiles indicate that a much larger share of the population has a small value than you thought (e.g., are holding one credit card where you thought most people are holding at least two)?

If this analysis identifies missing values coded with a particular number (e.g., 0 or 999), consider rerunning Step 3 also for these types of missing values in order to better understand their structure and implications for bias.

Anomaly Detection with Machine Learning

Manual anomaly detection has the big advantage that it draws on the entire knowledge of the data scientist—for example, the insight that it's possible to have a negative deposit balance on a checking account (so-called overdraft) but not to have a negative amount of children. It has, however, also a serious number of limitations—for example, a judgmental review of outputs can miss serious issues, often a data scientist can manually review only a prioritized set of variables, and the manual approach struggles with certain types of situations, in particular time-series data (e.g., usage patterns over time) and situations where only the combination of specific values of several variables is unusual (e.g., if you are trying to model digital identity fraud, neither having a scanned ID indicating the applicant to be female nor the image analysis of a real-time selfie indicating the presence of a beard would be unusual. However, the combination of an applicant sporting a hipster beard and using a woman's ID would be such an exceptional event that you probably would want to double-click this case to understand what's going on—it may be a very low-skill fraudster or an indication that the mapping of IDs into image files in your database has been corrupted).

Machine learning techniques can be extremely powerful in detecting such types of anomalies and I therefore recommend using them in addition to a manual review of the most important variables identified in Step 3. I am, however, a lot more hesitant in recommending a specific technique for two reasons: First of all, different data scientists have different areas of expertise, and while some might be very fluent in developing, say, an autoencoder in order to identify anomalies with a neural network, other data scientists I have worked with were a lot more at ease with the k-nearest neighbor concept. Second, some of these techniques are a lot younger than the other concepts I discuss in this book, and therefore still subject to a lot of new research, and there are also ever more powerful software packages coming to market that execute advanced machine learning algorithms to detect anomalies; therefore what is great today might appear rather suboptimal tomorrow.

For data scientists who are at the beginning of exploring these techniques, however, I recommend a three-step approach that uses relatively familiar concepts:

- First, conduct a k-means clustering of the observations in your sample using your favorite criterion to determine the number of clusters (e.g., the average silhouette,[3] one of those metrics that can be evaluated automatically).

- Then, measure the smallest Mahalanobis distance[4] of each observation to any cluster center.

- Finally, sort observations first by nearest cluster and then by distance to that cluster (from largest to smallest) in order to manually review unusual cases.

Clustering identifies groups (clusters) of observations that are "similar." Similarity is measured by distance to each other, with each variable used as a distance measure in one dimension. This implies three important things:

- Each variable must be numerical; for this again you can use the mapping of categorical and missing values into Weights of Evidence from Step 2 (data leakage).

- It is critical that for each real-world "dimension" there is only one variable used in the clustering as otherwise this dimension gets additional weight for each correlated variable—a typical mistake in k-means clustering is that many variables measure "size" and as a result, the clustering procedure comes back with the eye-popping discovery that there are small, medium-sized, and large cases; again, the PCA is therefore a critical prerequisite.

---

[3]Leonard Kaufmann and Peter J. Rousseeuw, *Finding Groups in Data: An Introduction to Cluster Analysis*, Wiley-Interscience, 1990.
[4]The Mahalanobis distance is a scaled version of the Euclidean distance as it is normalized by the standard deviation of each variable.

- As each cluster represents a large group of cases, you should expect that the front line is familiar with this type of situation; when discussing the data with experts and other people from real life, you therefore can pick one or a handful of cases from the center of each cluster (i.e., those cases with the *shortest* distance) to illustrate the *archetypes* of cases you are distinguishing and using to define what is "normal."

For time-series data, a similar approach can be applied, only that you can choose to define your sample as a time window (like in a moving average calculation). While an outlier would be a single data point along the time series that is far away from the cluster center, a different type of issue is present if a unit of observation (e.g., the person behind a record) moves further and further away from the cluster over time.

As you will be working down your way from the "weirdest" to the "least weird" case in reviewing the cases that are far away from any cluster center, the question is naturally when to stop. If you review the farthest three to five cases of a cluster and don't find any issue with them, you might want to conclude that there is at least no overwhelming evidence for harmful outliers; whenever you do find a problem, see if you can identify other observations with the same issue and consider putting all of them aside before continuing your investigation in order to maximize the number of *different* issues you can identify within a short period of time. And do note that the assessment of whether a case is normal or not may require sending it to a front-line expert.

Step 5: Correlation Analysis with Protected Variables

So far, we have taken a holistic approach to identify seeds for all kinds of biases in the data. However, often there will be a concern about a very particular bias, such as discrimination against a particular race or gender. In this step, we will detect to what extent such a bias is embedded in the data. This requires, however, that the protected attribute (e.g., race) is actually known for each observation. In reality, this attribute might not have been collected (and possibly even be prohibited from collecting) precisely because it is protected, making this particular analysis unfortunately impossible.

In order to run this analysis, I recommend running a t-test on each variable in the sample (this includes both the dependent variable and the weights of evidence for all independent variables), splitting the sample by a binary indicator variable for the protected class (e.g., the gender flag), and comparing the t-values.

First of all, we want to know if the *dependent* variable shows a significant difference between our protected class and the rest of the population—and if so, if the difference strikes us as material. If yes, we have a problem insofar as we should expect the algorithm to replicate this difference in real-world outcomes (i.e., there is a real-world bias present).

Second, we compare the t-values for all *other* variables with the t-value we obtained for the dependent variables. If the t-test is significant but the t-values are equal or smaller (i.e., any correlation is less or equally significant as the one we established for the outcome variable), the situation is contained in the sense that the dependent variables may help the algorithm to replicate the real-life bias but at least won't amplify it. However, if the t-value of a feature is larger than our benchmark, then we are dealing with a variable that could act as a relatively pure proxy for the attribute that represents the cause of an unwanted bias (e.g., gender) and thus help the algorithm to even amplify it.

Whenever the t-test is significant, critical design questions arise—for example, should the variable be dropped, should the variable be kept but the bias be addressed through a management overlay, or does a careful analysis of consequences suggest that the variable could be kept in the algorithm without any further adjustment?

In my experience, this bivariate diagnostic approach (comparing only two variables at a time) is sufficient. However, sometimes only a clever combination of multiple variables "gives away" the protected attribute—especially if the dataset includes computed features where the protected attribute might have been an input. If you want to make sure to detect such instances as well, run a linear regression (even for binary outcomes) with the protected variable as well as the explanatory variables you selected in Step 3 above (data structure) with the PCA (however, exclude all variables you already have flagged as problematic; otherwise they will come up again and muddle the assessment of the other variables) and examine the Variance Inflation Factors (VIF). My rule of thumb is that if the protected variable has a VIF of 2 or more, multicollinearity is present and the protected variable can well be approximated by a linear combination of other variables. The culprits can be detected by similar VIF values, and as you start dropping one or more of them, you can discover how many variables would need to be excluded in order to remove any hidden clue for the protected attribute.

Finally (to state what must be obvious to 99% of my readers), if the indicator variable is categorical with more than two classes (e.g., you distinguish three or more races), you will need to run ANOVA (Analysis of Variance) instead of the t-test.

Step 6: Visual Analysis

In our final step, we give our brain another chance to look at the data and call out odd situations—but this time using our brain's visual capabilities. For this, I plot a moving average of the dependent variable (on the Y-axis) against each continuous variable that was shortlisted in Step 3 (PCA)—with the dependent variable being the log-odds ratio for binary outcomes, its logarithm for approximately logarithmically distributed variables (in particular for amounts, such as income), and the simple mean for all others.

For ordinal variables, it sometimes makes sense to treat them as continuous (e.g., if there are many different levels), otherwise bar charts can be considered as long as the number of observations per category is meaningful (rule of 100).

Smooth relationships (picture-book-like straight lines, gentle curves, even U-shapes) give me comfort that a variable captures some underlying truth. Zig-zags, by contrast, attract my attention because they sometimes can point straight to a problem—for example, once a graph plotting the log-odds ratio for credit defaults against the number of continuous months for which a customer had made a payment showed a curious spike at 3.5! The odd value of the independent variable (you can have 3 or 4 but not 3.5 months of payment) gave away that here a junior analyst had conducted a thoughtless missing value imputation, putting in the sample average (hence the calculation of a value of roughly 3.5) for customers with missing payment history even though that segment had an exceptionally high default rate. If I hadn't caught it, this mistake could have caused a serious bias against customers who have made payments for exactly 3 or 4 months (as the algorithm might have treated them as high risk).

A second benefit of this visual inspection is that it transitions the data scientist's work from data engineering to model development as it also can generate insights for necessary transformations that can be addressed in feature generation.

Summary

In this chapter, we compiled a routine for a data X-ray that aims to be both comprehensive and efficient in raising warning flags for potential biases in the data. We structured the routine into six steps that build upon each other:

- The first step scans for completeness of the sample, in particular by assessing sample-level metrics and confirming that critical data fields do not have missing values.

- The second step aims to detect data leakage by a) detecting suspiciously high predictive power of individual variables and b) catching hindsight information encoded through missing values.

- The third step makes further diagnostic and remedial work a lot more efficient by extracting the structure of both missing values and predictive variables.

- The fourth step detects anomalies, leveraging a manual approach as well as machine learning.

- The fifth step specifically scans for seeds for biases against a protected class (e.g., Martians).

- The sixth step visualizes the data through cross-tabulations as a second opportunity for your brain to cry foul if it catches a glimpse of a potential anomaly.

By now, you will be intimately familiar with your data and many of the particular issues of the modeling problem at hand. In light of this, you now can make an informed decision about one of the most fundamental model design questions—namely which modeling approach to choose, and in particular whether to use machine learning or more manual techniques. This will be the focus of the next chapter.

When to Use Machine Learning

Just how the invention of computers and the Internet has fundamentally changed our world, machine learning is suddenly enabling analytics to be almost everywhere. Where such rapid change occurs, we humans are of course also prone to exuberance, even hype, and we sometimes need to take a step back and take a deep breath in order to keep things in perspective.

When I took a marketing class at University of Giessen in 1995, the professor used a chart from the 1960s that illustrated the idea of product life cycles with different types of fabric. Cotton was plotted at the dying end of its lifecycle while nylon was said to be at the start of a blistering career. Unless you are like my professor (who might have been away from Earth for a few decades, maybe studying Martians?), you must have noticed that this enthusiasm for nylon was overblown and short-lived: cotton was back in vogue already in the 70s and the use of nylon is limited to specific products (e.g., parachutes and winter clothing) where its advantages outweigh its limitations.

Keep the fate of nylon in mind when observing the current hype around machine learning. I will acknowledge machine learning's important benefits in a moment, and I am convinced that it is here to stay and will have a much broader set of uses than just pantyhose and winter jackets. However, don't

© Tobias Baer 2019

T. Baer, *Understand, Manage, and Prevent Algorithmic Bias*,
https://doi.org/10.1007/978-1-4842-4885-0_20

forget that machine learning—and neural networks in particular—emulate the working of the *subconscious* human brain, which is the brain's animalistic part, the same pattern-based, effortless, ultra-fast decision engine that lions use to hunt and that makes dogs good guard dogs (who are basically cute anomaly-detection machines with a loud bark). However, when nature designed humans, it upped its game and added a whole new capability which we call *logical thinking*. Logical thinking increased human energy consumption by a whopping 25% but that was more than outweighed by the increase in performance, or what we call intelligence. *Artificial* intelligence therefore at this point is designed to be more an intelligent poodle than a data scientist with PhD. I absolutely believe that every household would benefit from having a dog and would never claim that my ability to catch flying objects is anywhere as good as my parents' dog's aptitude. When it comes to developing algorithms, however, I still believe that an artisanal approach in many cases can produce the better models.

Machine learning's value proposition is mostly that it is fast and cheap. "Data mining" therefore, in my mind, has the appeal of fast food. Just like few people would argue that fast food tastes better than three-Michelin-starred cuisine, and your dining preference comes down mostly to your budget of time and money, I believe that machine learning comes in whenever you don't have the time to follow a more artisanal approach or lack, indeed, the necessary "cooking" skills (which is not at all surprising for certain types of data such as images).

"Artisanal" is not commonly used in the context of data science but after noticing that the best ice cream in town often is advertised as "artisanal" (and never as "machine produced"), I have come to believe that the concept of artisanship captures very well the value a skilled data scientist adds by constantly examining the data she works with and modifying her approach in order to circumnavigate specific limitations and potholes appearing in the data, thus deftly outmaneuvering harmful biases.

But haven't you heard that machine learning models are so much more predictive, too? In my observation, many contests where the machine learning model comes out as more predictive really can be explained by the data scientists building an artisanal challenger model not having enough time—and that gap might be a year for some types of really "big data" but it also may be just a week or two (many benchmarking exercises are really rushed). And these comparisons rarely acknowledge if the higher predictive power of the machine learning model came at the cost of a bias.[1] In fact, the best algorithmic cooks steal many of machine learning's best tricks, just as gourmet cuisine often is a high-tech affair using a whole machine park with torches, grinders, vacuum pumps, and much more. We will look at this in the next chapter. In this chapter, however, I would like to discuss the fundamental choice of whether a given modeling problem can be safely solved by machine learning or requires a more artisanal approach in order to manage algorithmic bias.

When we pit machine learning against an artisanal approach where the data scientist goes through the model development process in a much more manual fashion, infusing her business judgment and context knowledge into all modeling decisions, we can use the same cost-benefit-framework that we used in Chapter 12 to decide between an algorithm and judgmental decision-making. The artisanal approach will add both substantial labor cost and a significant delay in availability of the algorithm; machine learning, on the other hand, *can* introduce a substantial amount of business risk that an algorithmic bias occurs and causes financial and non-financial harm.

In order to quantify this business risk from algorithmic bias, three aspects need to be considered:

- First, we need to establish which biases might be present. For this, we have systematically scanned the data for evidence in Chapter 19.

- Next, we need to identify the specific actions a data scientist following an artisanal approach would take to deal with these biases but that would be omitted by a rapid, automated machine learning approach. Chapter 18 provides the background for this assessment and thus allows us to define the exposure to algorithmic bias that using machine learning for a particular modeling problem would create.

- Finally, we need to assess the severity of losses that could arise from this exposure to algorithmic bias, as discussed in Chapter 13.

---

[1] This is well illustrated by a hackathon I once staged where teams from all over the world built a credit score. In the development sample, the winning machine learning algorithm had a performance almost double of that of a logistic regression by a team that declared "robustness" their primary objective—but that advantage came crashing down to a paltry 2 Gini points for out-of-time validation, and upon closer inspection we realized that the machine learning model had engaged in "red lining," a practice that is illegal in the US and heavily discriminates against various groups of people including many blacks.

In practice, I have found that machine learning is advocated in particular if:

- **Speed** is paramount—if an algorithm is obsolete by the time an artisanal data scientist has crafted it, it is machine learning or nothing. Algorithms protecting from credit card transaction fraud and cyber security incidents are good examples of this category.

- **Insights** are low—if neither the data scientist nor the rest of the organization has much contextual insight and therefore no case can be made that an artisanal approach would offer any advantage over a machine learning model, the decision is probably down to machine learning or no algorithm at all (as discussed in Chapter 12). Organizations that just started to use algorithms very recently and hence possess little expertise in data science (and rely on software instead that automatically applied machine learning to their decision problems) may find themselves in this category, although they should ask whether this is where they want to remain in the long term.

- **Economic benefits** are small—if the business problem simply does not pay for an elaborate artisanal approach, again the decision is down to either machine learning or no algorithm at all.

On the flipside, an *artisanal* approach is inevitable if the economic downside of machine learning is grave and becomes particularly attractive if the amount of insight, time, and money available to build a fabulous algorithm is substantial.

Modeling problems involving very big or difficult-to-process data—for example, unstructured data such as text messages across multiple channels, complex time series data such as transactions with rich and possibly unstructured metadata, voice and image data—can pose modeling challenges that cannot be addressed in a "pedestrian" approach, just as the human brain relies on the poodle in its subconscious for tasks such as vision, language, and falling in love. Here true team work can be called for where artisanal data science and machine learning complement each other, just as a gentlemen with a sparkling subconscious has learned to "watch his tongue" and keep the unwanted, impulsive utterances of his subconscious at bay. This will be the subject of the next chapter.

Summary

This chapter contrasted machine learning with an artisanal approach in their respective ability to deal with algorithmic bias and derived implications for when each approach appears best suited. In particular, you learned that:

- We are most concerned about using machine learning if we have detected evidence that our algorithm will be **prone** to a bias that poses a **material** business risk and that can **only** be properly addressed by an artisanal approach.

- On the other hand, we may not be able to **afford** an artisanal approach if **speed** is paramount, our organization has few or no **insights** to treat the bias, or the **economic benefit** of the algorithm is insufficient to justify an artisanal development.

- If neither machine learning nor an artisanal approach is viable, using no algorithm at all may be the best option.

- In other situations—notably if we are dealing with very **big** or **complex data**—we want to pursue a hybrid approach that uses both machine learning and artisanal techniques.

In the next chapter, we will discuss what such a hybrid approach could look like.

How to Marry Machine Learning with Traditional Methods

When given a choice between keeping a cake and eating it too, I always try to find a way to do both. And I did succeed with that in the realm of machine learning!

You may have been surprised, maybe even upset, by the negative tone I took on machine learning in the previous chapter. This is not for lack of respect and admiration for machine learning—all I wanted to do was open your eyes to the limitations of machine learning. Giving machine learning tools to a data scientist who is not skilled in managing the risks of algorithmic bias can be as

© Tobias Baer 2019
T. Baer, *Understand, Manage, and Prevent Algorithmic Bias,*
https://doi.org/10.1007/978-1-4842-4885-0_21

dangerous as giving a Porsche to a novice driver. And given what we learned about the overconfidence bias, you hopefully will agree that I must shout really loud in order for my warnings to have any chance of being heeded by all readers!

The truth is, machine learning does give you a lot of power. And often the best algorithms live in the best of all worlds—an artisanal model carefully crafted by a watchful data scientist who employs a wide range of machine learning techniques as mere tools in her work. In this chapter, I want to introduce four specific techniques to incorporate machine learning techniques into artisanal model designs that allow for plenty of oversight by data scientists in order to prevent algorithmic bias:

- Feature-level machine learning

- Segmentation informed by machine learning

- Shift effects informed by machine learning

- Second opinion provided by machine learning

In the following sections, I will briefly explain each of them.

Feature-Level Machine Learning

Machine learning has a particular advantage over other techniques with very granular (and hence big) data, such as a situation where for each unit of observation (e.g., a patient or loan applicant) there are a large number of transactions, such as readings from a Continuous Glucose Monitoring system or credit and debit card transactions. It also has distinct disadvantages, such as a tendency to overfit (i.e., become unstable) if categorical variables have very rare categories or if there is a hindsight bias in the data. Why then not just limit the use of machine learning to specific features that are derived specifically from those data sources where machine learning is at its best?

If you build a stand-alone estimate from a particular data input using machine learning, you have a complex and hopefully highly predictive feature that you now can carefully embed in an artisanal equation—which means that you can apply any amount of modifications or constraints to safeguard algorithms from biases. In fact, this approach even opens up the possibility to use federated machine learning—an approach where the data used to develop the algorithm sits on distributed machines (e.g., many different users' mobile phones or Internet of Things refrigerators) and is never combined to one large database available to the data scientist's scrutiny; the approach estimates a stand-alone algorithm on each device and then only sends the algorithm itself to a central server that uses the aggregation of all algorithms to continuously come up with an optimized version that is distributed back to all devices.

How to keep such features at bay depends on the complexity of your chosen approach. An approach called *genetic algorithms* will create and test all kind of variable transformations and send you the best transformations it found; these often are still sufficiently transparent for a subject matter expert to judge whether the transformation makes sense and is safe from a bias perspective.

Other approaches will render the feature a black box. For example, consider a recruiting process for branch sales staff via video chat. Machine-learning-powered video analytics could measure what percent of time the applicant smiles—a probably useful indicator of an applicant's ability to build rapport with a customer, especially if the algorithm is able to distinguish fake smiles (where only one facial muscle moves—the one that can be consciously operated) from real smiles (which require two muscles to contract, one of which cannot be manipulated and hence truly reflects an empathetic emotional state). Here you cannot inspect the algorithm itself—instead, the presence of algorithmic bias in this feature needs to be established through the kind of analyses discussed in Chapter 19.

Let's now assume that back-testing has revealed that this algorithm works much better for Zeta Reticulans than for Martians—as a result, roughly half of the time that a Martian smiles, the algorithm fails to pick it up, giving Martians therefore a systematically lower friendliness score.

A data scientist becoming aware of this issue thanks to the X-ray she ran on the complex smiling feature (e.g., she might have observed a correlation between propensity to smile and race) now could solve the problem by converting the original smile feature (which measures percent of total talk time) to a rank variable. How can this eliminate racial bias? If the rank is calculated within a race, the "top 20 percent of smilers" will always contain both 20% of all Martians and 20% of all Zeta Reticulans, even if your machine learning algorithm claims that the most smiling Martian smiles less than half as much as the most smiling Zeta Reticulan.

Furthermore, if a sudden problem arises during day-to-day operations (e.g., your model monitoring reveals that the machine learning algorithm calculating your smiling feature has a huge bias in favor of people wearing clothes in warm desert and sunset hues, which popped up as the latest fashion trend), there is a stop-gap solution available to switch off this single feature without halting the entire algorithm (which is much harder to do if everything is baked into a single black box).

Segmentation

Another source of out-performance of machine learning algorithms over artisanally derived ones is the ability to detect subsegments that require a different set of predictors. Many artisanal workhorses such as logistic regression apply the same set of predictors to everyone, and it is often difficult for a data scientist to notice that there is a subsegment requiring a totally different approach (and hence a separate model).

In order to have the best of two worlds, I start by building both an artisanal model and a machine learning challenger model. I then calculate for each observation in my sample the estimation error for each model, and from that I derive the difference in error between the two models. A positive difference implies that for this observation, the machine learning model was better, a negative that the artisanal model was better.

Now you can run a CHAID tree with your PCA-prioritized shortlist of predictors (see Step 3 discussed in Chapter 19) to predict the difference in errors. Find the end nodes with the largest positive error difference (i.e., those where the average error of the artisanal model is much larger than the average error of the machine learning model) and trace back the variables and cutoffs that defined these subsegments. Do these subsegments make business sense? Are they a proxy for something else (possibly even a proxy for a segment defined by a variable not included in the modeling dataset)?

Classic examples coming out of such an analysis could be self-employed customers contained within a retail credit card sample where most customers are salaried, or a sizeable segment with missing credit bureau information. As always, discussing these results with the front-line can yield invaluable insights—sometimes the CHAID tree only approximates what should be the "correct" segment definition based on business insight.

Note that sometimes the segmentation amounts to the reflection of an external bias—for example, if in a highly discriminatory environment Martians tend not to be admitted to universities, you might realize that the CHAID tree's recommendation is to build a separate model for Martians because the whole set of features relating to the specifics of the applicant's university education only works for Zeta Reticulans. You see how you might very quickly get into some very tricky tradeoffs—but the whole advantage of this hybrid approach is that as the artisanal data scientist, you are firmly in the driver's seat to decide how to deal with this pattern in the data.

Once you have decided on the one or two segments that require a fundamentally different model, you can build separate artisanal algorithms for them. The result of this is stunning—often enough, through this technique I achieve a predictive power that is not just equivalent but *higher* than that of the machine learning benchmarking model (e.g., for binary outcomes, the outperformance may be 1-2 Gini points, which for some uses can be a lot of money—if your credit portfolio has an annual loss of $500 million, reducing that even by just a few percentage points would be enough for you to eat in a gourmet restaurant with 3 Michelin stars every night for the rest of your life...).

Shift Effects

Another source of insight by machine learning algorithms often missed by artisanal approaches is interaction effects; situations where only the combination of multiple attributes has meaning. For example, if your records indicate that the customer is female but the voice interacting with your automated voice response unit sounds male, you have a high likelihood of a fraudster impersonating your customer. In this case, there is a signal that should adjust the probability of fraud estimate upward. The perfect artisanal model would capture that signal but otherwise could still use the same set of predictive variables as for any other case in the population. Here the creation of subsegments (e.g., separate fraud models for female and male customers) would introduce unnecessary complexity (and effort)—instead, the shift effect technique simply adds additional variables to the artisanal model.

The by far easiest (and often sufficient) approach is to add a binary indicator (so-called *dummy*) for each shift effect. The model could be "patched up" by introducing a new variable that is 1 if the customer on record is female and the voice sounds male and 0 otherwise. You could also consider more complex adjustments, such as introducing an *interaction effect* of customer's gender and a "maleness" score of the voice ("tuning down" the warning signal if the sound of the voice is rather inconclusive).

The trickiest to identify (but often very powerful) opportunity lies in normalizing independent variables by dividing them through a contextual benchmark. For example, when I built a model to predict revenue of small businesses in an emerging market, I used predictors such as credit card receipts, electricity consumption, and floor space. The model had a couple of biases because in some industries, sales per square meter were particularly high, while in rural areas, credit cards were a lot less common than in cities. I therefore normalized the predictors by the median of their peers (e.g., rural versus urban pharmacies) and obtained a much more powerful (and equitable) model.

Just like the segmentation approach, the shift effect technique can also raise the performance of the hybrid model above that of the benchmark machine learning model. And it goes without saying that both techniques (segmentation and shift effects) can even be combined, giving pure machine learning a run for its money!

The Second Opinion Technique

When you compare the errors of an artisanal model and a machine learning model case by case, you will realize that the outperformance of the machine learning model only realizes itself *on average*—the number of cases where the machine learning model's estimate is worse than the artisanal model's one often is almost as large as the number of cases where the machine learning model does better.

A natural interpretation of this situation would be to conclude that cases where the two models disagree are in some way exceptional and therefore would benefit from the expertise of a human. The second opinion approach thus runs a machine learning model in parallel to an artisanal model and flags cases where the two have a large discrepancy for manual review. The manual review often can be greatly improved by having rules to flag the likely source of the discrepancy (e.g., by flagging attributes of the case that may be an anomaly—in many cases, all that is needed for the two models to align is for the human reviewer to adjust some input data) and by prescribing a framework or even specific steps for the manual review such as manually collecting specific additional information. In fact, in many situations I have created a full-fledged separate *qualitative* scorecard that not only systematically collects maybe 10-25 additional data points through the human reviewer but also actively eliminates judgmental bias by what I call psychological guardrails.

Strictly speaking, flagging a portion of cases for manual review by comparing two competing models is not a modeling technique. However, by including this in my list I want to reiterate my belief that the ultimate objective of the data scientist is to optimize a decision-problem, and the ideal architecture of the decisioning process may very well include steps outside of a statistical algorithm. Often the data scientist is uniquely positioned to advise business owners of such a possibility and create tremendous value by advocating it.

Summary

In this chapter, you learned that data scientists can indeed have the best of two worlds by using machine learning to extract valuable insights from the data while still using artisanal techniques to keep biases out of the model. Key take-aways are:

- Where big and complex data can only be harnessed with machine learning, rather than building one big black box model, you could consider using machine learning to build a set of complex features (typically using just a particular data source or set of data fields for each feature, giving it a tightly defined business meaning).

- When a benchmark machine learning model performs better than your artisanal model, you can use a CHAID tree to understand the types of cases that drive this outperformance.

- If you find that the outperformance of the machine learning benchmark model comes from specific subsegments for which your artisanal model is inappropriate, you can consider building separate models for such subsegments. Often the problem is really concentrated in just one or two such segments.

- If, by contrast, the outperformance comes from interaction effects, you can capture these effects through additional variables (e.g., indicator variables) that you incorporate in your artisanal model.

- Artisanal models enhanced with machine learning in this way often perform better than the machine learning benchmark model.

- If, however, even the artisanal approach does not succeed in removing a bias or the aforementioned techniques fail to replicate the insights garnered by the machine learning benchmark model, you should also consider revisiting the overall decision process architecture; it may be best to run artisanal and machine learning models in parallel and to have cases with strongly conflicting predictions of the two models reviewed by a human.

I disproved the adage that you cannot keep a cake and eat it, too. There is another piece of wisdom, though, that still holds: there's no free lunch—so you must pay for your miracle cake. The hybrid approaches laid out here require time because they are essentially manual. In some situations, however, that time for manual model tuning simply is not available—and this is nowhere as acute as in the case of self-improving machine learning algorithms. By the time a data scientist has updated an artisanal model, a self-improving machine learning algorithm has already gone through several new generations, making the data scientist's work obsolete before it even is finished.

In the next chapter, we therefore will discuss how best to keep biases out of self-improving machine learning algorithms.

How to Prevent Bias in Self-Improving Models

One of the greatest advantages of machine learning is that models can build and update themselves without any human intervention, enabling them to respond to structural changes at the fastest possible pace. The very context requiring such self-improving algorithms (the fast change of the environment in which they operate) is also the source of a heightened risk of biases affecting the algorithm, be it self-reinforcing feedback loops like we experienced in the context of social media (Chapter 11) or new data that might enable the algorithm to develop a bias against a protected class.

In this chapter, I therefore will describe a couple of specific tools that could help keeping algorithmic bias in check in self-improving machine learning models. The applicability and effectiveness of each technique depends highly on the context and therefore my intention is to give you useful pointers for your customized model design and not to describe a "one size fits all" solution.

© Tobias Baer 2019
T. Baer, *Understand, Manage, and Prevent Algorithmic Bias*,
https://doi.org/10.1007/978-1-4842-4885-0_22

In order to safely operate self-improving machine learning models, I suggest considering three elements:

- Model mechanics

- An "emergency brake" that prevents an updated version of the model to go live if a warning signal for a material bias flashes

- And regular manual review of inputs, characteristics, and outputs of self-improving models in production (this process does not prevent a biased model from going live but it limits downside risks by attempting to catch problems quickly)

Finally we will discuss how to deal with *real time* machine learning, which is the fastest-paced version of self-improving models.

Model Mechanics

In this category, the model design, the data engineering, and the algorithm estimating the model require your particular attention.

The **model design** could limit the risk of biases by "boxing" self-improving machine learning, as these two examples illustrate:

- By building *separate models* for Martians and Zeta Reticulans the self-improving algorithms focus on better ranking for people within their peer group but it is impossible for the overall decision engine to develop a bias against one of the groups.

- Self-improving ML algorithms could also merely be *features* in a stable, artisanally derived model. For example, a resume screening algorithm could be an artisanal logistic regression that consists of discrete self-improving machine learning-based scores for aspects such as "technical knowledge," "goal achievement," and "people leadership."

The **data engineering** part should import all the best practices discussed in Chapter 18 into the mechanics of the self-improving algorithm. This implies in particular:

- The scripts feeding fresh data into the algorithm need to correctly apply all required exclusions (e.g., immaterial defaults) and data cleaning steps (e.g., dealing with outliers).

- Consider using only data from continuous randomized trials for refreshing the algorithm (as opposed to biased feedback reflecting choices made by prior versions of the algorithm).

- When deciding the time window from which you pull data for the model's refresh, consider adapting the length of the window for the frequency with which a particular type of event or feature occurs (e.g., rather than always looking at just the last seven days of Internet usage, recognize that while one feature (e.g., a feature measuring to what extent a person seems to explore multiple options before making a decision) might find more than enough data points in a seven-day window, another feature (e.g., a feature mapping specific web sites or search terms into a risk profile) might require a year's worth of data because it deals with categorical variables where many values are quite rare and hence seven days' worth of data will leave the mapping of most terms statistically inconclusive (and hence open the door wide for biases).

- Correctly recognize unknown values—there is a real danger in new categories popping up because they often end up accidentally being interpreted as something else. This is because code mapping categorical data into a risk index or bucket often includes an "everything else" clause (e.g., an algorithm automatically separating vegetarian from non-vegetarian menu items might first label items containing chicken, beef, veal, lamb, venison, or deer as "meat," then search for keywords indicating fish or seafood, and label everything else as "vegetable," promptly treating your new guinea pig dish as a potato). The correct treatment would be to define an "I don't know" category and possibly even route such cases into an exception handling routine (i.e., to not make an automated decision without human intervention).

- Consider embedding automated bias detection routines into your script so that if a new bias has arisen in the data, the script can sound an alarm bell and automated updates could possibly even be suspended.

And for the script **estimating the model**, I recommend in particular the following two safeguards:

- Limit what kind of features the self-improving algorithm can create and test. There are scripts that will create thousands of transformations including, say, the cube of the cosine of the account number, and if such an esoteric feature never came up during the initial model development, its sudden appearance in the model would be dubious and is more likely to introduce a new bias by overfitting the data than constituting a break-through insight into esoteric numerology;

- Code a thorough automated validation routine that as a minimum should include an in-period hold-out sample and an out-of-time validation sample but also could include a special "exam" sample of cases that specifically tests for biases and metrics for the stability of estimates. In my own work, I often use a procedure that generates thousands of candidate models, sorts them by predictive power or business impact, and then validates one after the other until it finds one that passes all tests. Given my obsession with model robustness, rarely did the first model pass all exams posed to it, but most of the time the model chosen did not perform much worse.

The design of such constraints on the self-improving machine learning algorithm of course faces the usual tradeoff between squeezing out maximum predictive power (which can translate into real business value) and robustness that reduces the risk of harmful biases. If you want to have the best of both worlds, you could also consider running a system where the self-improving algorithm in production is heavily constrained while in parallel, you run a "freestyle" self-improving model purely for benchmarking, and if that algorithm finds a material outperformance by considering the cube of the cosine of your credit card number, an alert pops up on your desk.

Considerations in Designing an Emergency Brake

The emergency brake complements the automated validation routine we discussed above in the context of the model estimation; it focuses on the continuous flow of model inputs and model outputs.

When designing an emergency brake, two decisions stand out: what triggers it and what is the implication.

The line between triggers of the "emergency brake" and what would only raise a concern in the post-implementation manual review of a self-improved model is a subjective one and depends on risk appetite as well as practical considerations. The metrics must lend themselves to an objective criterion for the trigger and the number of false alarms must be acceptable.

Calibrating the tradeoff between safety and keeping the number of false alarms manageable may require a bit of trial and error; if the majority of updates of the model equation trigger a manual review, you don't really have a self-improving algorithm anymore. In order to arrive at appropriate emergency brakes, I recommend a three-step approach:

1. Compile the entire list of metrics you want to monitor post-implementation (see the next subsection) and define for each metric your risk appetite (i.e., how much change or deterioration in performance you are willing to accept).

2. Back-test your hypothesized metrics on a simulated run of your self-improving algorithm on historical data and track how frequently each metric would have rang an alarm bell; for those metrics causing too frequently a false alarm, consider whether you want to change the trigger point for the emergency brake or drop the metric altogether.

3. Launch the self-improving algorithm and keep track how often each metric triggers a false alarm; reconsider any metric that does this too often.

You also need to decide what to do if the emergency brake is activated: do you want to freeze (i.e., keep) the previous version of model until the self-improved version is manually assessed, or do you want to suspend the decision process altogether? This will depend on whether the environment has experienced a material structural change or if just something has gone awry with your latest iteration of the algorithm. The self-improving algorithm could assess this by also testing the previous version of itself on the fresh data; if that triggers the emergency brake as well, it does not appear prudent to continue any algorithm for automated decision-making at all and the automated decisioning should be suspended.

The idea of emergency brakes also could be used in regulating self-improving algorithms. For example, right now there is a problem with machine learning for medical uses because the Food and Drug Administration treats every updated version of an algorithm as a new "device" that requires a new approval, to the tune of a two-year-long administrative process. A more trigger-based approach to safety regulations, by contrast, is known from countries with mandatory safety inspections for vehicles, where the permission to drive the car is given for maybe a couple of years during which only material alterations of the car would trigger an immediate requirement of a new safety inspection (but not the installation of a bigger rear mirror or the attachment of a bumper sticker advertising this book).

Model Monitoring

Which metrics should you monitor? I recommend tracking four forward-looking aspects of the model:

- Population profile
- Model output (predictions)
- Model attributes
- Out-of-sample validation performance

The population profile could be measured by a distribution analysis of important attributes of a case, such as key model inputs and maybe a few other attributes that are actively tracked by business users (e.g., banks usually track the distribution of loan applicants by credit bureau score even if their credit scoring algorithm uses a lot more granular inputs than that summary score); it is an effective way to detect structural changes in the population.

You also could cluster the "base population" (e.g., the sample on which the initial model was developed) and measure the percent of cases in the most recent sample that are far away from the center of all original clusters; if that share rises, it is likely that a new segment or profile is entering the population. While this is exactly the type of situation for which the self-updating algorithm is designed, it also means that an additional manual validation could be called for because you may not have the right input data to predict outcomes for this new segment.

In doing this, it is important that the reference population is "frozen" (e.g., the sample the data scientist used when the initial version of the self-improving algorithm was developed and the various parameters and constraints were decided) because otherwise the algorithm might be caught "asleep" (or even be manipulated), especially if anomaly detection is its main purpose (e.g., in the context of fraud and cyber security). Just as bacteria can become immune against an antibiotic if they initially ingest just a small quantity of the antibiotic, an algorithm comparing today's population with yesterday's may not sound the alarm if the quantity of a particular type of "unusual" transactions starts very small and then rises slowly from day to day.

Model output refers to the forward-looking metric discussed in Chapter 15; if you suddenly see the approval rate or average predicted income wander off, chances are that your model has caught a bias bug.

Model attributes refer to our discussion about XAI (explainable machine learning) and monitoring self-improving algorithms in Chapter 15.

Out-of-sample validation calculates two sets of analyses:

- The backward-looking metrics discussed in Chapter 15 calculated on some historic validation sample for which outcome labels are available

- The comparison of the distribution of model outputs between the new version of the algorithm and some earlier reference version

The manual review of these metrics complements the thresholds for automated triggers. Material changes in the population profile or the model output could trigger an emergency brake; material changes in model attributes or significant issues in out-of-sample validation already should trigger the model estimation procedure itself to reject the algorithm.

That implies that the manual review of these metrics would focus more on situations where either a metric is in the "amber" zone (timidly ringing a warning bell without crossing a red line) or there is a deteriorating *trend* in these metrics. We want to know whether based on this data, the human reviewer might piece together an indeed alarming trend or the insight how some change in the real world is slowly eroding the analytic approach taken by the decision system, which may require you to fundamentally rethink which data should be used for making the decision at hand.

In addition, also the backward-looking metrics on rank-ordering capability and calibration discussed in Chapter 15 should be monitored. If outcomes are available by the time the metrics discussed here are reviewed, they could be included in the same report; otherwise, this would be a separate monitoring process due to the time delay. The monitoring of these metrics is not any different from traditional models.

Last but not least, if you are facing a situation where there is a particular concern about a specific bias (e.g., whether Martians are discriminated against), you could include the direct measurement of correlation between model outputs and the protected variables in this report, as discussed in Step 5 of Chapter 19.

Real-Time Machine Learning

Before closing this chapter, I want to briefly comment on real-time machine learning. While other types of self-improving machine learning algorithms go through discrete versions that are developed at regular time intervals (or triggered by some event), real-time machine learning continuously produces new algorithms. There can be as many new versions as there are new cases streaming into the system.

Real-time machine learning is not applicable everywhere. It requires also real-time labeling (i.e., defining the outcome of each case such as the good/bad indicator), and whenever labeling requires some human interaction or otherwise a de facto batch process (e.g., a "default" happens if an account remains unpaid for 90 days, which happens when the clock strikes midnight or an accounting system goes through end-of-day processing), these processes will define the earliest moment when an algorithm might self-improve.

However, what do you do if real-time machine learning does happen, such as in search optimization where click-through data streams in as fast as new search queries? Here real-time emergency brakes are neither possible nor necessary. They (probably) are not possible because of speed considerations (the algorithms must be estimated in lightning speed), but they are also unnecessary because of the stability bias of algorithms. A single observation is as much able to materially change an algorithm as a bird pushing on a super tanker is able to turn the ship around.

Which is to say unable, unless the bird carries a jet pack—or the new data point is an extreme leverage point. A single extreme outlier does have the power to bias an algorithm and therefore introduces the risk of not only accidental bias but also willful manipulation, and I can easily imagine applications where there are plenty of economic incentives to do so, such as when trying to outsmart automated trading algorithms (those robots that trade in stock and foreign exchange markets). And what a single data point cannot achieve maybe could be achieved by a salvo of manipulated data points…

Protecting real-time machine learning from biases therefore requires a two-pronged approach. On the one hand, it is absolutely critical to keep tabs on the input data—automated data cleaning that floors and caps values or otherwise guards against outliers should be combined with dedicated monitoring of anomalies in the input data, especially if there is anyone out there who might have an interest in causing harm to your organization.

On the other hand, you need to run a background process that in frequent, regular intervals evaluates the triggers for the emergency brake and intervenes if a problem occurs (which includes, of course, problems with the input data).

Summary

Self-improving algorithms are essentially developed by auto-pilot. In order to set up this auto-pilot to prevent biases, a couple of practices stand out:

- Model design can limit the danger of algorithmic bias by intentionally boxing self-improving machine learning algorithms within an overarching model design.

- Data engineering needs to carefully embed all relevant techniques to prevent algorithmic bias from slipping in through the automated data feeds.

- The automated model estimation procedure can limit the risk of biases by imposing constraints on automated feature generation and automatically validating each new version of a model.

- An emergency brake automatically monitors the population profile and model outputs and either reverts to a safe, earlier version of the model or halts the entire automated decision process if a set of carefully chosen and calibrated triggers rings an alarm.

- Regular (ex post) manual monitoring complements the automated safeguards against bias.

In this and the previous four chapters, we have discussed how as a data scientist, you can systematically integrate techniques to fight algorithmic bias into your own model development. If you are managing a team of data scientists, however, knowing the techniques is not enough—you also need to think about how to institutionalize these techniques. This will be the subject of the final chapter of this book.

How to Institutionalize Debiasing

The first thing I learned as a consultant is that it is easy to have good ideas—the real challenge is in implementing them! Having read 22 chapters of this book already, your head hopefully is brimming with good ideas on how to fight algorithmic bias. But how do you make it happen, especially if you, say, oversee a couple of hundred data scientists who are chasing deadlines and prove to be human by exhibiting their fair share of overconfidence bias?

In this chapter, I will introduce seven specific steps to institutionalizing the debiasing practices discussed in this book into your organization. These steps address:

- Data flow and warehousing
- Standards and templates
- Materiality framework
- Calibration
- Model validation
- Model monitoring
- Continuous generation of unbiased data

© Tobias Baer 2019
T. Baer, *Understand, Manage, and Prevent Algorithmic Bias*,
https://doi.org/10.1007/978-1-4842-4885-0_23

Data Flow and Warehousing

We may be living in the age of machine learning—but we also could say that we live in the age of the Chief Data Officer (CDO), a new role that has been adopted by many organizations with the objective to secure access to "the new oil," namely data. Throughout the book, I have stressed the importance of data—hence keeping models free of biases often also requires investments in data quality, primarily by increasing control and automation of data flows (e.g., to ensure completeness of data and a "single truth" in the sense that for a given attribute of a given object, an organization's IT systems should offer a single (and true) answer—it is surprising how many different answers some banks' IT systems offer when asking how much money a given customer has borrowed).

Additional, less obvious activities for a Chief Data Officer could be:

- Establish a foolproof way to access historical data without leakage;

- Scout and provide enterprise-wide access to external data that can complement or replace biased internal datasets;

- Establish standard routines to detect biases in datasets and alert businesses of the presence of biases;

- Increase the organization's data literacy by publishing not only an authoritative data dictionary (thus preventing misunderstandings about the meaning of data that could cause biases) but also a guide on known data limitations and other hidden sources of biases.

Standards and Templates

Every data scientist has her own approach to modeling that reflects both the techniques she is most comfortable with and deep convictions; changing such practices is hard. I found the combination of two tools most effective in changing practices nevertheless:

- Create a document that defines "standards" for the area you are responsible for—when I built a large modeling team for the first time, I humbly called my collection of best practices "our bible" and tirelessly trained each class of new joiners on these standards as this is the easiest moment to mold a data scientist's approach.

- Create templates that imply a certain approach simply by prescribing certain outputs for specific steps in a process.

Table 18-1 in Chapter 18 is an example of such a template—mandating a particular format for model documentation can go a long way in enforcing a set of practices. I also created templates for specific steps in the model development process such as Tables 19-1 and 19-2 in Chapter 19 or the formatting of the Principal Component Analysis.

Sadly, many templates sit unused on some shelf collecting dust. In order to actually get them applied, you probably need to make use of two additional techniques:

- Managers must stubbornly enforce the use of templates top-down, including yourself—simply refuse to look at anything that doesn't use the prescribed template, and make sure that managers on lower levels in the hierarchy (e.g., your team leaders) do the same. If gentle prodding is not enough, consider establishing "percent of analyses using the correct template" as a metric that is officially tracked and, if peer pressure proves insufficient, incorporate it in the KPIs (key performance indicators driving compensation) of team leaders and other managers.

- Have your team create off-the-shelf scripts that automate the generation of the standard output as much as is feasible (reports such as what I showcased in Tables 19-1 and 19-2) and prudent (you do not want to provide off-the-shelf scripts where you want your team to actually do some heavy thinking!). Convenience is another strong lever to influence behavior, and it obviously feels a lot better than coercion!

Materiality Framework

The ancient temple of Delphi had a Doric saying carved in stone: "μηδεν αγαν," nothing in excess—Greek philosophers referred to the same principle as "the golden mean." Heed this ancient wisdom when raising your organization's awareness for biases. As some quantum of bias is present in most real life phenomena, an excessive concern for biases might paralyze your team when unbiased data suddenly seems to be all but impossible to come by.

Where you draw the line between immaterial biases and reason for concern will depend on many factors, including legal requirements for your industry and specific types of decisions, cultural sensitivities in the culture and market in which you are operating, and your and your organization's risk appetite and convictions. As you start your crusade against algorithmic biases, you may want to start by clearly articulating where you see the biggest issues and why (as this will both focus initial efforts and create a benchmark for what is a "problem") and over time try to create a more general articulation what kind of biases are unacceptable versus what kind of situations are considered immaterial.

Calibration

As discussed in Chapter 4, there are two important calibration steps involved in translating the output of an algorithm into a decision: the calibration of the *central tendency* and the calibration of *decision rules*.

Central tendency is the average of what you want to predict—statistically this relates to the constant term c of the equation you encountered in Chapter 3. An often overlooked detail is that this calibration should reflect the average outcome you will encounter *in the future*—it is a classic stability bias to calibrate to the average observed in the past instead. For example, when I calibrate a credit scorecard, I need to calibrate it to the portfolio default rate I will encounter in the, say, next 12 months. Would the data scientist know?

Well, unless the data scientist has a side job as a clairvoyant, he wouldn't. The business people would be the closest to the question but might be blinded by heavy interest biases. Therefore I always recommend to my clients to make this calibration decision a well-defined committee decision (e.g., in banks, a credit committee that represents both the business and the risk management sides would be naturally predisposed for this) and ideally to explicitly debias the process (e.g., create strong anchor points by running multiple forecast models using different techniques such as a vintage model, a momentum model using roll rates, and a time series model using macroeconomic factors).

Decision rules define where to draw the line between yes (which may mean anything from a new credit card to early release from prison) and no—and this act of drawing a line could include specific adjustments to compensate for algorithmic biases, as discussed in Chapter 16. Just as in the case of the central tendency, the way to facilitate such adjustments fairly and prudently is to elevate the calibration of decision rules to an explicit management decision as opposed to burying it in a black-box algorithm that only provides a magic yes/no decision.

Model Validation

As discussed in Chapter 18, there is tremendous value in an independent validation of a data scientist's work—it has the potential to bust everything from a casual bias caused by an oversight of a fatigued data scientist to a deep-rooted bias of said data scientist covered under a two-feet-thick layer of overconfidence. Early on in my career managing a team of data scientists I had the utopian dream of informal peer review but I had to conclude that effective peer review is as likely to happen as your cat begging you for a daily shower.[1]

Formal validation has the added benefit that it also can be a formidable power to enforce standards and templates—if your organization's governance rules require the validation team to sign off on a new algorithm before it can be implemented, and if the validation team will refuse to even look at any model where the documentation does not follow the template you have defined, you have won that battle at least!

I have to admit that this idea of formal validation isn't my own—some clever banking regulator beat me on that one, and nowadays in the financial industry some form of formal model validation is required in most jurisdictions. In fact, what is missing today is wider knowledge of classic scriptures in general and inscriptions of the temple of Delphi in particular, as many banks today are plagued by what I would consider *excessive* validation practices, as briefly discussed in Chapter 18.

The situation is very different in the corporate world where many organizations just are starting to make widespread use of algorithms and hence validation practices are nascent at best. I should note that the value of formal model validation practices for fighting algorithmic bias is almost a fringe benefit—faulty models often can have substantial, if not catastrophic, business impact, so if there is a business case for developing an algorithm, most likely there is a strong business case for formal validation, too.

Model Monitoring

A piece of wisdom that seems to be missing on the temple of Delphi (and instead is ascribed by an unverifiable source to a fifteenth century writer called Gabriel Bell)[2] is "you get what you pay for." In the case of corporations, this means that if you really want something to happen, just pay someone to do it. And one thing that really deserves hiring someone for is model monitoring.

In this respect, banks have been the frontrunner, but even in this industry a few years ago I caught a bank that hadn't monitored a model that had been in use for over seven years. That model had developed a bias simply due to inflation—it had a feature comparing an applicant's income to the "average" income, and if that average is taken from a time when a haircut still cost just

---

[1]In case you are not familiar with cats, try giving a cat a shower *just once*. Or if you don't want to be torn into pieces, just take my word that it won't happen.
[2]www.charlesmccain.com/pdf/Who-said-it-Original.pdf

25 cents (I'm exaggerating just *slightly*), every single applicant appears to be the richer brother of William A. Rockefeller Jr.

This anecdote points at a more fundamental challenge: many companies including banks actually struggle to keep track of all the models they use. While some "high profile" models are formally validated and monitored by central functions, nobody in the headquarters might even be aware of a model used by a small personal loan business that is operated by a subsidiary of the Papua New Guinea branch. Only recently have banking regulators started to formally require banks to assemble a "model inventory" which surfaced all kind of interesting things.

Not only did this reveal that some models were used "under the radar" and without proper monitoring, but it also shed light on a different problem: some models were used opportunistically (e.g., they were used to approve loans that looked like they would be approved by the model but they were bypassed for loan applications that looked like they would be rejected), thus introducing a new decision bias on its own.

To ensure that biases are reliably detected as soon as they appear somewhere in the organization—be it in the data that flows into its algorithms or the outputs coming out from them—a dedicated person or team to monitor models is therefore critical. Their job should include an explicit mandate to catalogue all models used in the organization and to ensure that every model is monitored unless it is immaterial or doing so is not economical.

A best practice is that explicit rules translate serious issues appearing in the ongoing monitoring into mandatory remedial action, which could include anything from a review of the situation by a data scientist to a suspension of the algorithm. And in order to avoid overconfident biased humans from bypassing algorithms where they disagree, the monitoring should cover the entire decision process, including human overrides and decisions made without consulting the algorithm.

Continuous Generation of Unbiased Data

And at the risk of sounding like a broken record, I want to reiterate one last time the importance of striving for a continuous flow of fresh, unbiased data to test and refine algorithms. In my experience, organizations that excel in this don't do so because of the efforts of a lone data scientist—they have embraced testing and learning as a culture, and proposing a new product or other business idea without a plan for generating such data elicits as puzzled a look by colleagues as if you serve a Bavarian a white sausage (Weißwurst) without a generous helping of sweet mustard.

To truly eliminate algorithmic biases, it therefore is important that senior managers set the tone by understanding where algorithmic biases could occur, by asking which techniques their colleagues have deployed to keep biases at bay whenever an algorithm is involved in something, and where applicable by asking about the approach for collecting or generating unbiased data going forward. This routinely will entail the need for explicit experiments or other data collection efforts as well as a budget for paying for this.

Summary

In this last chapter, you explored the challenge of embedding debiasing in a large organization and learned about seven important practices:

- Complete and unbiased data requires investments in automating the flow of data and the warehousing of a "single truth." The Chief Data Officer is a new role that increasingly champions and coordinates these efforts.

- Standards and templates are effective ways to train staff and a fabulous basis (but not sufficient) for enforcing a consistent approach to preventing algorithmic biases.

- A materiality framework is important to prevent paralysis through excessive bias-phobia.

- Calibration is an important step in translating algorithmic outputs into decisions that should be elevated to a management decision as opposed to being part of a data scientist's black-box design.

- Model validation is invaluable in countering biases of data scientists and enforcing standards, but only if it is formally mandated.

- Model monitoring complements model validation but likewise requires formalization in order to happen reliably.

- The continuous generation of unbiased data is such a critical part of the solution to algorithmic bias that it should be embraced by top management as a credo and part of the corporate culture.

With this, we conclude our journey through the world of algorithmic bias. We neither have been able to find a magic bullet against it, nor have we found much reassurance that algorithmic biases will ever disappear. However, I hope that you have been able to develop a thorough understanding of the many sources of algorithmic biases and that you have learned concrete and practical steps that you can take to detect, manage, and prevent algorithmic biases, whatever your role and level of expertise.

In parting, I am sharing one last list with you—it's a tongue-in-cheek list of "10 commandments" I sometimes share with data scientists in particular when working on credit scorecards (although they are much more broadly applicable). Humor often travels much faster and farther than stern warnings, so feel free to share this in your organization as part of your crusade against algorithmic biases.

THE 10 COMMANDMENTS OF MODEL DEVELOPMENT

1. Thou shall read and apply this handbook, as it is good and the result of many lessons learned the hard way.

2. Thou shall always start and finish your modeling work with understanding how the business user is using the scorecard, as the scorecard is just a tool, and the wrong tool in the wrong man's hand can be a deadly weapon.

3. Thou shall always let light shine on your work. The business user is the light, so you let the light shine by sharing almost daily neatly synthesized outputs with your project counterparts on the business side.

4. Thou shall always look at the data, as he who doesn't look at things with his eyes is blind, and the blind at times are not aware of lethal dangers.

5. Thou shall never use a variable or feature you do not understand or cannot explain, as the devil may be hiding in it.

6. Thou shall not have highly correlated variables in the same model as they are the tools with which the devil does his black magic.

7. Thou shall always keep everything—the model structure, the features, the modeling techniques—as simple as possible without losing significant performance as the devil is hiding behind complexity.

8. Thou shall keep features and scores continuous wherever you can as steps are cliffs, and many a man has died by falling from a cliff.

9. Thou shall always thoroughly validate a model on fresh, unused data, as the data scientist not validating a model with fresh data is merely reading tea leaves arranged by the devil.

10. Thou shall always think of big chunks of cases where your model is weak (e.g., because of systematically missing values or different behavioral profile) as the Lord of Segmentation says that thou shall have other models or modules where your first model doesn't work.

Index

I

A

B

C

© Tobias Baer 2019
T. Baer, *Understand, Manage, and Prevent Algorithmic Bias*,
https://doi.org/10.1007/978-1-4842-4885-0

Printed in the United States
By Bookmasters